Achieving Quantitative Literacy

An Urgent Challenge for Higher Education

Library of Congress Catalog Card Number 2003115950

ISBN 0-88385-816-9

Printed in the United States of America

Current printing (last digit):
10 9 8 7 6 5 4 3 2 1

Achieving Quantitative Literacy

An Urgent Challenge for Higher Education

Lynn Arthur Steen

Published and Distributed by

The Mathematical Association of America

National Forum
Quantitative Literacy: Why Numeracy Matters for Schools and Colleges

Steering Committee

Richelle (Rikki) Blair, Professor of Mathematics, Lakeland Community College, Kirtland, Ohio

Peter Ewell, Senior Associate, National Center for Higher Education Management Systems

Daniel L. Goroff, Professor of the Practice of Mathematics, Harvard University

Ronald J. Henry, Provost and Vice President for Academic Affairs, Georgia State University

Deborah Hughes Hallett, Professor of Mathematics, University of Arizona

Jeanne L. Narum, Director of the Independent Colleges Office of Project Kaleidoscope

Richard L. Scheaffer, Professor Emeritus of Statistics, University of Florida

Janis Somerville, Senior Associate, National Association of System Heads

Forum Organizers

SUSAN L. GANTER, Associate Professor of Mathematical Sciences, Clemson University

BERNARD L. MADISON, Visiting Mathematician, Mathematical Association of America

ROBERT ORRILL, Executive Director, National Council on Education and the Disciplines (NCED)

LYNN ARTHUR STEEN, Professor of Mathematics, St. Olaf College, Northfield, MN

The MAA Notes Series, started in 1982, addresses a broad range of topics and themes of interest to all who are involved with undergraduate mathematics. The volumes in this series are readable, informative, and useful, and help the mathematical community keep up with developments of importance to mathematics.

MAA Notes

11. Keys to Improved Instruction by Teaching Assistants and Part-Time Instructors, *Committee on Teaching Assistants and Part-Time Instructors, Bettye Anne Case,* Editor.
13. Reshaping College Mathematics, *Committee on the Undergraduate Program in Mathematics, Lynn A. Steen,* Editor.
14. Mathematical Writing, by *Donald E. Knuth, Tracy Larrabee, and Paul M. Roberts.*
16. Using Writing to Teach Mathematics, *Andrew Sterrett,* Editor.
17. Priming the Calculus Pump: Innovations and Resources, *Committee on Calculus Reform and the First Two Years,* a subcomittee of the Committee on the Undergraduate Program in Mathematics, *Thomas W. Tucker,* Editor.
18. Models for Undergraduate Research in Mathematics, *Lester Senechal,* Editor.
19. Visualization in Teaching and Learning Mathematics, *Committee on Computers in Mathematics Education, Steve Cunningham and Walter S. Zimmermann,* Editors.
20. The Laboratory Approach to Teaching Calculus, *L. Carl Leinbach et al.,* Editors.
21. Perspectives on Contemporary Statistics, *David C. Hoaglin and David S. Moore,* Editors.
22. Heeding the Call for Change: Suggestions for Curricular Action, *Lynn A. Steen,* Editor.
24. Symbolic Computation in Undergraduate Mathematics Education, *Zaven A. Karian,* Editor.
25. The Concept of Function: Aspects of Epistemology and Pedagogy, *Guershon Harel and Ed Dubinsky,* Editors.
26. Statistics for the Twenty-First Century, *Florence and Sheldon Gordon,* Editors.
27. Resources for Calculus Collection, Volume 1: Learning by Discovery: A Lab Manual for Calculus, *Anita E. Solow,* Editor.
28. Resources for Calculus Collection, Volume 2: Calculus Problems for a New Century, *Robert Fraga,* Editor.
29. Resources for Calculus Collection, Volume 3: Applications of Calculus, *Philip Straffin*, Editor.

30. Resources for Calculus Collection, Volume 4: Problems for Student Investigation, *Michael B. Jackson and John R. Ramsay*, Editors.
31. Resources for Calculus Collection, Volume 5: Readings for Calculus, *Underwood Dudley*, Editor.
32. Essays in Humanistic Mathematics, *Alvin White,* Editor.
33. Research Issues in Undergraduate Mathematics Learning: Preliminary Analyses and Results, *James J. Kaput and Ed Dubinsky,* Editors.
34. In Eves' Circles, *Joby Milo Anthony*, Editor.
35. You're the Professor, What Next? Ideas and Resources for Preparing College Teachers, *The Committee on Preparation for College Teaching, Bettye Anne Case,* Editor.
36. Preparing for a New Calculus: Conference Proceedings, *Anita E. Solow,* Editor.
37. A Practical Guide to Cooperative Learning in Collegiate Mathematics, *Nancy L. Hagelgans, Barbara E. Reynolds, SDS, Keith Schwingendorf, Draga Vidakovic, Ed Dubinsky, Mazen Shahin, G. Joseph Wimbish, Jr.*
38. Models That Work: Case Studies in Effective Undergraduate Mathematics Programs, *Alan C. Tucker,* Editor.
39. Calculus: The Dynamics of Change, *CUPM Subcommittee on Calculus Reform and the First Two Years, A. Wayne Roberts,* Editor.
40. Vita Mathematica: Historical Research and Integration with Teaching, *Ronald Calinger,* Editor.
41. Geometry Turned On: Dynamic Software in Learning, Teaching, and Research, *James R. King and Doris Schattschneider,* Editors.
42. Resources for Teaching Linear Algebra, *David Carlson, Charles R. Johnson, David C. Lay, A. Duane Porter, Ann E. Watkins, William Watkins,* Editors.
43. Student Assessment in Calculus: A Report of the NSF Working Group on Assessment in Calculus, *Alan Schoenfeld,* Editor.
44. Readings in Cooperative Learning for Undergraduate Mathematics, *Ed Dubinsky, David Mathews, and Barbara E. Reynolds,* Editors.
45. Confronting the Core Curriculum: Considering Change in the Undergraduate Mathematics Major, *John A. Dossey,* Editor.
46. Women in Mathematics: Scaling the Heights, *Deborah Nolan,* Editor.
47. Exemplary Programs in Introductory College Mathematics: Innovative Programs Using Technology, *Susan Lenker,* Editor.
48. Writing in the Teaching and Learning of Mathematics, *John Meier and Thomas Rishel.*
49. Assessment Practices in Undergraduate Mathematics, *Bonnie Gold,* Editor.
50. Revolutions in Differential Equations: Exploring ODEs with Modern Technology, *Michael J. Kallaher,* Editor.
51. Using History to Teach Mathematics: An International Perspective, *Victor J. Katz,* Editor.
52. Teaching Statistics: Resources for Undergraduate Instructors, *Thomas L. Moore,* Editor.
53. Geometry at Work: Papers in Applied Geometry, *Catherine A. Gorini,* Editor.
54. Teaching First: A Guide for New Mathematicians, *Thomas W. Rishel.*

55. Cooperative Learning in Undergraduate Mathematics: Issues That Matter and Strategies That Work, *Elizabeth C. Rogers, Barbara E. Reynolds, Neil A. Davidson, and Anthony D. Thomas,* Editors.
56. Changing Calculus: A Report on Evaluation Efforts and National Impact from 1988 to 1998, *Susan L. Ganter.*
57. Learning to Teach and Teaching to Learn Mathematics: Resources for Professional Development, *Matthew Delong and Dale Winter.*
58. Fractals, Graphics, and Mathematics Education, *Benoit Mandelbrot and Michael Frame*, Editors.
59. Linear Algebra Gems: Assets for Undergraduate Mathematics, *David Carlson, Charles R. Johnson, David C. Lay, and A. Duane Porter,* Editors.
60. Innovations in Teaching Abstract Algebra, *Allen C. Hibbard and Ellen J. Maycock*, Editors.
61. Changing Core Mathematics, *Chris Arney and Donald Small,* Editors.
62. *Achieving Quantitative Literacy: An Urgent Challenge for Higher Education,* Lynn Arthur Steen.

MAA Service Center
P. O. Box 91112
Washington, DC 20090-1112
800-331-1622 fax: 301-206-9789

Foreword

Over the last two decades the mathematical community has become increasingly aware that it is important for our students, as well as in our own best interests, to engage in wide-ranging discussions of what and how we teach. *Achieving Quantitative Literacy* stands in this tradition. Quantitative literacy (QL) is one of today's hot topics. This is due, at least in part, to efforts in the past several years by Lynn Steen and his colleagues at the National Council on Education and the Disciplines (NCED). They have raised the visibility of QL and challenged the mathematical community to devote serious attention to it. The appearance of this volume in the *Notes* series of the Mathematical Association of America (MAA) is an important response to this challenge.

It is a fortunate coincidence that *Achieving Quantitative Literacy* appears at the same time as the most recent MAA curriculum report, *Undergraduate Programs and Courses in the Mathematical Sciences: CUPM Curriculum Guide 2004.* This *Guide* from the MAA Committee on the Undergraduate Program in Mathematics (CUPM) is imbued with the essence of the QL message: Mathematics is an exciting and central human activity and, as a consequence, departments of mathematics and mathematical sciences have an obligation to position themselves at the core of their institutions. This includes active leadership for quantitative literacy.

Quantitative literacy does not need to be taught only by mathematicians any more than effective writing needs to be taught only by English professors. Mathematicians are not necessarily the best prepared to teach it. But each mathematics department has a responsibility to nurture and shape a meaningful program in quantitative literacy. More than a responsibility, the challenge to create such a program presents an opportunity for mathematics to take its rightful place of influence and importance at the heart of undergraduate education.

David M. Bressoud, Chair
Committee on the Undergraduate Program
in Mathematics (CUPM)
Macalester College
February, 2004

Preface

Literacy in the information age requires the ability to comprehend information that is presented in words and numbers, in sentences and tables, in diagrams and graphs. Literacy experts describe this broad range of contemporary demands as *prose*, *document*, and *quantitative* literacy. Each dimension extends from primitive elements taught in primary school through the challenges of college-level study in subjects such as history, economics, and biology.

Students' experience with prose literacy extends through all subjects and all grades by means of reading and writing assignments. Curiously, however, the same is not true for document and quantitative literacy. To literacy experts, document literacy refers to reading charts and tables, quantitative literacy to interpreting and reasoning with numbers. As students move through secondary and tertiary education, the silos created by individual disciplines tend to ignore both document and quantitative literacy, leaving them orphans without obvious advocates. Even mathematics slights them, focusing its instruction not on the interpretation of data but on the procedures of algebra. In this report, we combine both under the label quantitative literacy (QL), or its near-synonym, numeracy

Notwithstanding the importance of quantitative literacy to health, politics, work, and personal finance, in our discipline-dominated education system QL has neither an academic home nor an administrative promoter. To understand better the educational challenges presented by quantitative literacy, the National Council on Education and the Disciplines (NCED) initiated a national examination of issues surrounding QL, especially in the context of school and college studies.

As a starting point, NCED published *Mathematics and Democracy: The Case for Quantitative Literacy*, consisting of a case statement on numeracy in contemporary society followed by twelve responses.[1] Subsequently, to expand the conversation, the Mathematical Sciences Education Board (MSEB) hosted a national forum on quantitative literacy that was supported by NCED in cooperation with the Mathematical Association of America

(MAA). Proceedings of this Forum were subsequently published under the Forum's title: *Quantitative Literacy: Why Numeracy Matters for Schools and Colleges*;[2] this volume includes papers commissioned as background for the Forum, essays presented at the Forum, and selected reactions to the Forum.

The present report, *Achieving Quantitative Literacy*, provides a synopsis of major issues raised at the QL Forum supported by numerous quotations from the Forum, both short and long. Its purpose is to pervade undergraduate education with a consciousness of the importance of QL and thereby to create a national debate about the place of quantitative literacy in higher education. Reflecting the Forum itself, ideas and evidence come from many different fields and perspectives. Part I contains analyses, findings, and recommended responses; Part II offers supporting materials: dialogues on related topics, examples of QL assessments, and descriptions of programs in the new National Numeracy Network. Together, these different components also serve as an entrée to the lengthier papers published in the Forum proceedings.

Although the primary audience for this synopsis (as for the Forum proceedings) are postsecondary faculty, administrators, and members of boards of trustees, speakers at the Forum made clear the importance of a secondary audience as well: policy-oriented people who are trying to find ways to hold post-secondary education accountable for demonstrating that students have learned something of value. To both audiences, the report conveys several important messages:

- Quantitative literacy largely determines an individual's capacity to control his or her quality of life and to participate effectively in social decision-making.
- Educational policy and practice have fallen behind the rapidly changing data-oriented needs of modern society, and undergraduate education is the appropriate locus of leadership for making the necessary changes.
- The wall of ignorance between those who are quantitatively literate and those who are not can threaten democratic culture.
- Quantitative literacy is not about "basic skills" but rather, like reading and writing, is a demanding college-level learning expectation that cuts across the entire undergraduate curriculum.
- The current calculus-driven high school curriculum is unlikely to produce a quantitatively literate student population.

The findings and responses outlined in this report provide opportunities for addressing QL in many forums, including undergraduate curriculum and assessment (both in mathematics and across disciplines); faculty preparation and professional development; horizontal networking among campuses; and vertical K–16 articulation. Professional associations and funding agencies too must look to quantitative literacy as fertile territory for supporting important change in higher education.

Finally, an important word about references. Since this report is derived directly from the Forum proceedings referenced above, *Quantitative Literacy: Why Numeracy Matters for Schools and Colleges,* it is filled with quotations—short and long—from that volume. To make this summary readable, these appear without further citation as to their source. Other references, several dozen in all, are cited in endnotes.

Acknowledgements: Much of the credit for initiating, sustaining, and refining the quantitative literacy "initiative" goes to my colleagues Robert Orrill, Bernard Madison, and Susan Ganter. From his historian's perspective, Bob recognized the profound significance of quantitative literacy both for higher education and for society; he orchestrated the QL initiative and helped many educators recognize its importance. Bernie recruited authors and panelists for the QL Forum that forms the basis for this report and has carried the message personally to numerous higher education forums. Susan identified and supported campus QL-related initiatives both in her role as director of the National Numeracy Network and through leadership of the Curriculum Foundations project of MAA. Their work, and mine, has been strengthened by advice from a joint MAA, NCED, and MSEB steering committee whose members are named at the beginning of this report. In addition, MAA assisted in distribution of the two QL volumes published by NCED, and became publisher of this report, the third in this series. Throughout, the QL initiative has been financially supported by a generous grant from the Pew Charitable Trusts.

Lynn Arthur Steen
Northfield, Minnesota
August 2003

Contents

Part I

The Third R: Quantitative Literacy

> My Ph.D. is in mathematics; by most standards, I was very "well trained." Nonetheless, the mathematics education that I received was in many ways impoverished. —Alan Schoenfeld[3]

> QL advocates need to be very clear about what all students need to know and be able to do, starting with where it fits in the mathematics program. —Janice Somerville

In the 1920s the United States crossed the threshold into universal secondary education, thereby raising new questions about the role of education for employment, for engaged citizenship, and for democratic well being. Now, a century later, we are crossing a similar threshold into universal postsecondary education. Consequently, we must now ask many of the questions about higher education that were asked about high school in the 1920s. But our age is dominated by computers and data, not factory assembly lines. As society has become more complex, literacy has become more sophisticated. Plain old 'rithmetic, the original third R, is clearly no longer sufficient for today's world. Scarcely any issue facing society can be resolved without recourse to sophisticated quantitative analysis and argumentation. It is this uniquely modern blend of arithmetic with complex reasoning that we call quantitative literacy (QL) or numeracy.

As reading and writing—the first two Rs—are part of education from primary school through college and beyond, so must be quantitative literacy. Although the basic elements of reading and writing are part of the K–12 curriculum, continued growth in both is universally recognized as an essential aspect of college education. So it is with quantitative literacy: although most mathematical tools required for QL are taught in school, continued reflective experience with data throughout college deepens students' capacity to use these tools for productive lives and responsible citizenship.

Half a century ago, before computers inundated our discourse with quantitative data, only a fraction of high school graduates went on to higher education. This is no longer true. Indeed, college enrollments as a percent of 18–24 year olds have risen 10% a decade since the end of WW II (Figure 1) so that today as many students enroll in postsecondary education as in secondary school (Figure 2). For many reasons, the American public has made college the locus of expectations for leading a decent life. The widely recognized rewards of sophisticated quantitative reasoning that can be acquired most readily through college and graduate programs are surely among those reasons.

As enrollments in higher education have grown nearly four-fold during the last forty years (from approximately 4 to 15 million students),[4] the curriculum has remained relatively stagnant. With few exceptions, today's college students study a curriculum that was developed for different students in a different time and for a different purpose. The U.S. prides itself on offering the best postsecondary education in the world, especially suited for educating a political and technical elite (and one that is increasingly international). Our system of higher education was conceived at a time when 2% of the U.S. population went to college and its purpose then was to educate students to become a class apart. But life in the 21st century is more complex, so the fundamental education for all citizens that formerly was accomplished by the end of second-

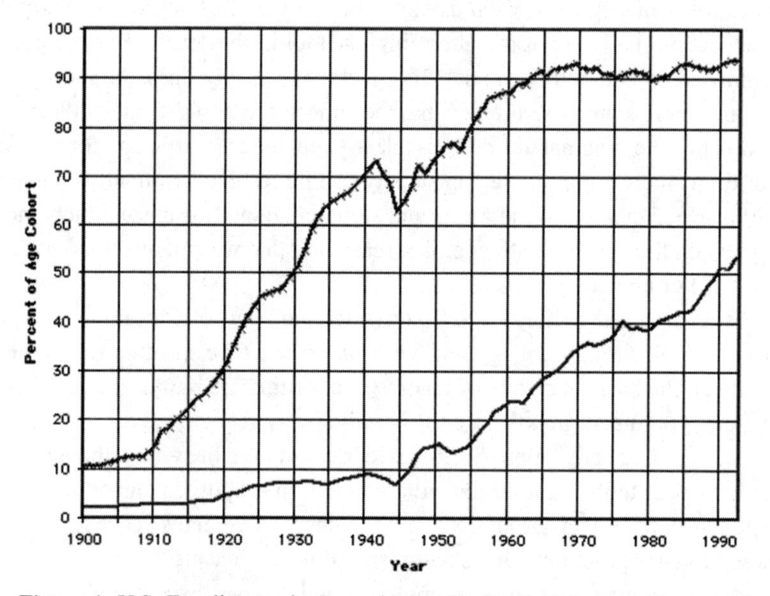

Figure 1. U.S. Enrollments in Secondary School and College as Percent of Age Cohorts, 1900–1992[5] [*Top:* Secondary School as % of 14–17 year olds. *Bottom:* College as % of 18–24 year olds]

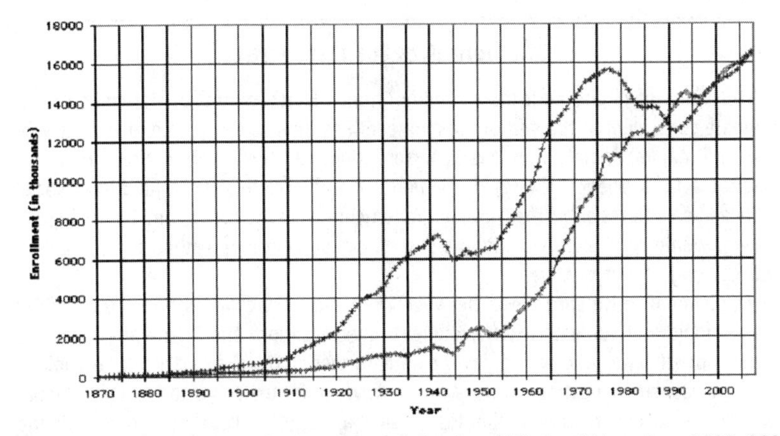

Figure 2. U.S. Enrollments in Secondary School and Higher Education, 1890–2005[6] [*Top*: Secondary School (in thousands). *Bottom*: College (in thousands)]

ary school now takes much longer—as the greater numbers going on to college testify. As many commentators have noted, college has replaced high school as the educational standard for a democratic society in the twenty-first century. However, the task of re-conceiving the structures of higher education for a democratic purpose—especially in fields that rely on quantitative and mathematical methods—has hardly begun.

Indeed, for reasons that will be expounded below, much the reverse has taken place. The curriculum in mathematics has gradually narrowed, forcing almost everyone through the bottleneck of calculus before expanding again into a stunning variety of powerful methods that undergird society's power brokers—scientists, financiers, cinematographers, pollsters, *et al.* Given that what counts as literacy now goes well beyond basic school-level skills—and notwithstanding many efforts to "reform" undergraduate mathematics to remedy perceived inadequacies of content or pedagogy—it may well be that with regard to a democratic conception of higher education it is undergraduate mathematics that is most out of date (see sidebar).

There is an irony in this that is worth noting. The very forces that now press colleges to address issues of quantitative literacy were created by colleges and universities in the first place. Changes in society that demand widespread quantitative literacy arose in large measure from innovations of college graduates seeking greater effectiveness and efficiencies in the fabric of life and work. But these same institutions failed to recognize that as their graduates changed the world of work and discourse, their curriculum needed to

Mathematics for Democracy

Historians of higher education identify Charles Eliot's reform of Harvard's undergraduate curriculum at the end of the nineteenth century as the source of today's subjects and majors. Prior to that, the "courses" students chose were entire curricula—classical, arts, or science. The rapid fragmentation of modern science (and later, social science) created irresistible pressure for greater choices in the curriculum. By freeing the undergraduate curriculum from its pre-scientific past, Eliot bequeathed to us what we now call majors and electives.

Empowered by Harvard's lead, between 1890 and 1920 colleges across the country created majors in subjects to offer students specialization, and "general education" courses for students who would never pursue further study in the particular discipline. This new "general education" became the conveyance for the democratic ideal of higher education—to ensure that graduates of all majors left college equipped with the breadth of general knowledge and skills necessary for full participation in democratic society.

An examination of college catalogues reveals that mathematics is an exception to this story. Unlike other subjects, educationally speaking, mathematics was not invented in the late nineteenth century. It is the one classical survivor in the modern curriculum. Prior to 1900, mathematics had a long and highly regarded presence in the curriculum, dating all the way back to the medieval quadrivium. Who was Eliot to change that?

So as mathematics joined other subjects in developing a major—clearly evident in changes in college curricula between 1900 and 1925—it did not need to, nor did it, rethink in any significant way the courses offered to students who did not major in mathematics. These courses, titled variously higher algebra, conic sections, trigonometry, solid and analytic geometry survived Eliot's curricular revolution little changed, gradually morphing into what is now generally termed college algebra or precalculus. Unlike professors in other subjects, a college professor from 1850 would not find today's college algebra course at all unfamiliar.

Both the stability of today's college algebra course and its senseless use as the primary vehicle for general education in mathematics is undoubtedly due to this peculiar history: college algebra is a relic of a prior curricular philosophy that no longer exists. And in the century since Eliot, nothing new has succeeded in addressing Eliot's challenge: to provide college graduates with a mathematics appropriate for citizenship.

change in order to prepare the following generation for this changed society. Fortunately, but slowly, this disconnect is being addressed.

Many signs suggest that the movement to enhance the role of quantitative literacy in education is already gaining momentum. There is not as yet, and probably should never be, a single QL movement; but as annealing creates a new material by minor jiggling of many molecules, so many local movements by different actors may gradually realign the educational system to be more supportive of quantitative literacy. Evidence of agitation towards QL fills publications, projects, presentations, and the activities of diverse educational and professional organizations (see sidebar).

Well before QL became a topic of widespread conversation among educators, a public debate had been going on about the relationship between education and the health of civil society in America.[7] The sources of this agitation come from both the political right and left (e.g., Robert Putnam, Alan Wolf, and Jean Bethke Elshtain). Disappointingly, there seems to be little or no attention to QL in this conversation even though statistical and quantitative measures are intimately connected to both social policy and civic health. Nonetheless, at least a few QL projects on college campuses are attempting to build some linkages between quantitative literacy and the formation of public policy.

As QL helps draw mathematicians and statisticians into discussions with their colleagues in other disciplines, they will undoubtedly begin to see means of relating QL to other reform initiatives (e.g., teacher education, first year programs) or to powerful historical developments (e.g., educational accountability, or Robert Putnam's arguments about the failure of civic society). We may learn from these discussions not so much what QL is, but rather why it must be talked about whenever goals of undergraduate education are on the table.

Evidence of ferment about QL can be seen both in the United States and in other nations. Typically, innovative QL programs on college campuses involve faculty from several disciplines reflecting the critical cross-disciplinary nature of QL. Many organizations concerned about undergraduate education, including both funding agencies and international organizations, have recently recognized the importance of quantitative literacy.[8] The accompanying sidebar on "liberating literacy" by Mogens Niss, former secretary of the International Commission on Mathematics Instruction (ICMI), illustrates one stream of international thinking about the importance to democracy of citizens' deep capacity for quantitative literacy.

Campus movements towards QL often connect with or are part of other campus priorities such as first year programs, teacher preparation, assessment, calculus reform, or general education. For example, the Greater Expectations project of Association of American Colleges and Universities (AAC&U) documents ways in which higher education is becoming conscious of its own changing landscape, one feature of which is the need to better nurture quantitative competence.[9] Others speak of a "deep capacity for literacy and numeracy" as the bedrock on which solutions to the many problems facing higher education must rest. Quantitative literacy is of immense importance to many of the special

Quantitative Literacy Gains Momentum

(*A selection of recent meetings, presentations, and publications focused on QL.*)

Meetings

Project Kaleidoscope Faculty Workshop on Quantitative Literacy (Park City UT, July, 2001)

Workshop on College Algebra, Precalculus, and Quantitative Literacy (Crystal City, VA, October, 2001)

National Forum: Why QL Matters for School and College (Washington, DC, December, 2001)

Conference on College Algebra (United States Military Academy at West Point, NY, February, 2002)

Project Kaleidoscope Faculty workshop on Quantitative Literacy (Williamsburg, VA, June, 2002)

QL Workshop for Geoscience Faculty (Northfield, MN, July 2002)

Numeracy and Beyond, Pacific Institute for the Mathematical Sciences (Vancouver, BC, July 2003)

Panels and Presentations

Florida Commission on Higher Education (Tampa, FL, November 2001)

American Association for the Advancement of Science (AAAS) Annual Meeting (Boston, MA, February, 2002)

Quality in Undergraduate Education [QUE] (Reno, NV, March, 2002)

AAC&U Greater Expectations Conference (Washington, DC, February, 2002)

National Institute on Aging, (Bethesda, MD July, 2002)

MathFest (MAA Summer Meeting), Burlington, VT, August, 2002)

Joint Statistical Meetings (New York, NY, August, 2002)

National Council of Teachers of Mathematics (NCTM) Regional Meeting (Biloxi, MS, October 2002)

Northern California Council of Teachers of Mathematics (Monterey, CA, December 2002)

Contributed Paper Session, Joint Mathematics Meetings (Baltimore, MD, January 2003)

Publications

"Numeracy: The New Literacy for a Data-Drenched Society," *Educational Leadership*, October 1999. URL: tinyurl.com/4yth

"Reading, Writing, and Numeracy," *Liberal Education*, Spring 2000.

Mathematics and Democracy, National Council on Education and the Disciplines, May 2001.

"Reflections on Quantitative Literacy," *Project Kaleidoscope*, July 2001.

"Quantitative Literacy," *Education Week*, September 2001. URL: www.edweek.org/ew/ewstory.cfm?slug=01steen.h21&keywords=Steen

"Quantitative Literacy: Everybody's Orphan," *Focus*, September, 2001.

"In a World of Data, Statisticians Count." *AmStat Online*, September 2001. URL: www.amstat.org/publications/amstat_news/2001/pres09.html

"Achieving Quantitative Literacy: The Challenges." *AWIS Newsletter*, January 2002

"The Circumference of a Circle." *Education Week*, February, 2002.
URL: www.edweek.org/ew/newstory.cfm?slug=22packer.h21&keywords=arnold%20packer
"Educating for Numeracy: A Challenging Responsibility." *AMS Notices*, February 2002.
URL: www.ams.org/notices/200202/commentary.pdf
"Quantitative Literacy: Why Numeracy Matters for Schools and Colleges," *Focus*, February 2002. URL: www.maa.org/features/QL.html
"Quantitative Literacy and SIAM," *SIAM News*, April 2002.
URL: www-math.cudenver.edu/~wbriggs/qr/siam_news.html
"Teaching Quantitative Skills in the Geosciences" (website).
URL: dlesecommunity.carleton.edu/quantskills/index.html

"programs that enhance learning" that were identified by *US News and World Reports* in their annual rankings of colleges and universities.[10] In recognition that numeracy is increasingly as important as literacy, the National Survey of Student Engagement (NSSE) recently added a few questions about students' problem solving experiences to parallel their experiences in writing papers.[11]

For too long undergraduate education has either overlooked the transforming implications of our data-saturated world or else assumed without much reflection that the appropriate response is to encourage more students to study more mathematics. But as many will testify, valuable though mathematics is for many reasons, the standard mathematics curriculum was never designed to promote quantitative literacy. In many cases, required mathematics courses actually subvert QL by convincing students that mathematics is not good for anything in the real world.

Society too focuses its response naively on mathematics—notably on standards and high stakes tests in the K–12 arena. In this they are abetted by inattentive college and university mathematicians who tolerate (if not encourage) public anxiety about college admissions, which focus far too narrowly on calculus and its predecessor courses. The result is an "impoverished" quantitative education, even for those who go on to careers in the mathematical sciences. And for those who do not, it too often results in what John Allen Paulos and others have decried as "innumeracy."[12]

It is time for a change, time to recognize the distinctive quantitative requirements of universal education in the computer age. As participants in the QL Forum repeatedly argued, personal success in the new information economy requires a new set of problem-solving and behavioral skills that emphasize the flexible application of reasoning abilities. These skills involve sophisticated reasoning with elementary mathematics more often than elementary reasoning

Liberating Literacy

Traditionally, we tend to see the role of mathematical literacy in the shaping and maintenance of democracy as being to equip citizens with the prerequisites needed to involve themselves in issues of immediate societal significance. Such issues could be political, economic, environmental, or they could deal with infrastructure, transportation, population forecasts, choosing locations for schools or sports facilities, and so forth. Issues may also deal with matters closer to the individual like wages or salaries, rents and mortgages, child care, insurance or pension schemes, housing and building regulations, bank rates and charges, etc.

While all this is indeed essential to life in a democratic society, I believe that we should not confine the notion of democracy, nor the role of mathematical literacy in democracy, to matters like the ones just outlined. In order for democracy to prosper and flourish, we need citizens who are not only able to seek and judge information, to take a stance, to make a decision, and to act in such contexts. Democracy also needs citizens who can come to grips with how mankind perceives and understands the carrying constructions of the world, i.e. nature, society, culture, technology, and who have insight into the foundation and justification of those perceptions and that understanding. It is a democratic problem if large groups of people are unable to distinguish between astronomy and astrology, between scientific medicine and crystal healing, between psychology and spiritism, between descriptive and normative statements, between facts and hypotheses, between exactness and approximation, or do not know the beginnings and the ends of rationality, and so forth and so on.

The ability to navigate in such waters in a thoughtful, knowledgeable, and reflective way has sometimes been termed "liberating literacy" or "popular enlightenment." As mathematical literacy often is at the center of the ways in which mankind perceives and understands the world, mathematical literacy is also an essential component in liberating literacy and popular enlightenment. We should keep that in mind when shaping education for the pursuit of mathematical literacy in service of democracy.

— Mogens Niss

with sophisticated mathematics. They manifest the ability to make sense of real world situations and to make judgments grounded in data.

Although QL is no less important than reading and writing for productive living in a democratic society, it is rarely recognized as an explicit goal of education. In contrast to mathematics, which for centuries has held a privileged place in education, the role of QL as a fundamental literacy is a relatively recent phenomenon, being largely the product of widespread use of databases and computer modeling. It is not surprising, therefore, that there is as yet no established consensus about exactly what QL entails, its role in the curriculum, or where and by whom it should be taught.

There is, however, an emerging consensus on the importance of quantitative literacy and the need for focused effort to improve students' abilities in

Quantitative Literacy Goes to College

(Examples illustrating the variety of QL programs on college campuses)

Special courses targeted at quantitative literacy

Mathematical Thinking (Bloomsberg University)
Workshop Mathematics (Dickinson College)
Case Studies in Quantitative Reasoning (Mount Holyoke College)
Foundations of Scientific Inquiry (New York University)
Core Competencies in Mathematics (Northern Illinois University)
Patterns of Problem Solving (State University of New York at Stony Brook)
Quantitative Literacy for the Life Sciences (University of Tennessee)
Quantitative Problem Solving (Valparaiso University)
Contemporary Applications of Mathematics (Virginia Commonwealth University)

Programs that introduce QL across the curriculum

Quantitative Reasoning Program (DePauw University)
Quantitative Reasoning Across the Curriculum (Hollins College)
Connecting Science and Mathematics in General Education (Hope College)
Quantitative Methods for Public Policy (Macalester College)
Integrating Science and Mathematics into a Humanities Curriculum (Trinity College)
Mathematics Across the Curriculum (University of Nevada, Reno)
Assessment of Quantitative Literacy (Virginia Commonwealth University)

this area. This consensus is reflected in five findings of the NCED/MAA/MSEB Forum on Quantitative Literacy:

Finding 1. **Preparation:** Most students finish their education ill prepared for the quantitative demands of informed living.

Finding 2. **Awareness:** The increasing importance of quantitative literacy is not sufficiently recognized by the public or by educational, political, and policy leaders.

Finding 3. **Benchmarks:** The lack of agreement on QL expectations at different levels of education makes it difficult to establish effective programs for QL education.

Finding 4. **Assessment:** QL is largely absent from our current systems of assessment and accountability.

Finding 5. **Professional Support:** Faculty in all disciplines need significant professional support in order for them to enhance the role of quantitative literacy in their courses.

Evidence for these findings is summarized in the pages that follow, together with recommended responses.

Quantitative Literacy in Higher Education

> In no part of U.S. education are the problems caused by disunity (or lack of articulation) greater than they are in mathematics.
>
> —Bernard Madison

> Although the mathematical foundation of quantitative literacy is laid in middle school, literacy can only be developed by a continued, coordinated effort throughout high school and college.
>
> — Deborah Hughes Hallett

> If educators cannot fulfill their economic mission to help our youth ... become successful workers, they also will fail in their cultural and political missions to create good neighbors and good citizens.
>
> —Anthony Carnevale

> I am convinced that we have to invent mechanisms of quality assurance that look at how courses are being taught as well as simply whether the content of the courses seems right.
>
> — Russell Edgerton

Calls for improved quantitative reasoning of college graduates come from many quarters (e.g., administrators, business leaders, political commentators), but generally these are not the most influential voices in discussions of higher education. On campuses, curricular talk tends to be dominated by disciplines. However, many signs suggest that higher education professionals are becoming conscious of the need to adapt to a changing and more complex landscape on which cultural understanding, civic responsibilities, diversity and religious perspectives contend. Two widely known examples are cited in sidebars: *Greater Expectations* and *Project Kaleidoscope.*

For reasons cited throughout this report and discussed at the Forum on which it is based, quantitative literacy will be among the more important issues debated on campuses as they reposition their programs to educate students for the twenty-first century. Colleges and universities typically include

Greater Expectations

A project of the Association of American Colleges and Universities, the Greater Expectations' National Panel advocates a contemporary vision for learning that addresses the "multiple hopes Americans hold for college education" and the role that higher learning plays in "creating a just democracy, cooperation among diverse peoples, and a sustainable world." The Panel's report urges college to ensure that their graduates are "empowered through the mastery of intellectual and practical skills, informed by knowledge about the natural and social worlds and about forms of inquiry basic to these studies, and responsible for their personal actions and for civic values." Concerned that the college experience is a "revolving door" for millions of students, Greater Expectations advocates a dramatic reorganization of undergraduate education to ensure that all college aspirants receive not just access to college, but an "education of lasting value." Further details, including the full report, can be found at www.greaterexpectations.org/.

quantitative or analytical competence as one of the expectations they promise to help students achieve, and increasing numbers of degree programs require students to demonstrate some level of quantitative competence.

Higher education is being urged by its many constituencies to get ahead of the curve,[13] to re-examine entrenched practices and become proactive in adapting to a changing environment. Empirical studies of students' course-taking patterns show that the core curriculum has been "quantified" without any intentionality or systematic plan.[14] Programs that ignore this reality will become diluted and leave students less well prepared than those that engage fully the challenge of quantitative literacy.

When faculty do address the role of QL, whether as part of a core curriculum discussion or in support of different majors and programs, there is always

Project Kaleidoscope

A project of the Independent Colleges Office in Washington, DC, Project Kaleidoscope (PKAL) is an informal national alliance working to build and sustain strong learning environments for undergraduates in mathematics, engineering, and science. The driving goals of PKAL are to equip teams of faculty and administrators for leadership in reform at the local level. To achieve those goals, PKAL employs a kaleidoscopic approach that gives attention to all facets of the undergraduate learning environment—from quality of faculty to character of facilities, from design of the curriculum to the shape of institutional cultures and budgets. Recent PKAL summer workshops have featured strands on "Quantitative Literacy: Everybody's Orphan" that bring together teams of faculty from different campuses to work together to shape QL projects at the local level. Further details can be found at www.pkal.org/.

considerable disagreement about what QL means. Even more debated are who should be responsible for QL instruction and what priority QL should have among the many goals of higher education. Inevitably, mathematicians will be at the center of this debate and—for good or ill—their actions will greatly influence the outcome. With no discipline naturally exercising leadership, there is neither an insistent nor a consistent call to make quantitative literacy a priority of education at the college level. And by not paying attention to QL in their own programs, colleges convey strong messages to secondary schools that QL really doesn't matter much. Until higher education pays attention to QL, high schools will not.

Making QL an Institutional Objective

It may be necessary to focus our discussion of quantitative reasoning separately on three populations of students: the enhancement of quantitative reasoning for all students, for those students entering K–12 classrooms as teachers, and as numeracy is applied in the disciplines and in professional programs. I see each of these tasks as different and requiring separate conversations and strategies. "For all students" speaks to issues of society and civic responsibility—those abilities that are necessary for an educated citizen to make informed decisions. Separate attention to preparing future teachers recognizes the fundamental need for quantitative approaches in all areas of the K–12 curriculum and not only by teachers of mathematics. The third area of attention, quantitative reasoning in all professional programs, is essential if we are to move to increasing levels of sophistication in application.

At least from my perspective, quantitative reasoning will not be accomplished if it is restricted to general education or allocated solely to mathematics. It must be an overall goal of institutions, defined as learning outcomes for graduating students and assessed directly as part of the evaluation of major programs and what those fields are demanding.

Gone are the days when we would require a course and not build on it in major programs. Why not require programs to set explicit goals for quantitative reasoning, develop plans for enhancing quantitative skills beyond basic mathematics and statistics courses required and then report on the results as part of their program assessments? This approach would contribute greatly to a dialogue and ultimately to significant cooperation between mathematics, statistics, and the disciplines (or professional programs) in achieving overall goals for students.

Appealing to the disciplines for their assistance and attention to quantitative reasoning is essential. It would also be transformative, building what we should expect in coordinated learning outcomes across an institution and in the process redefining a rich, liberal education. A true liberal education should demand the consideration of quantitative reasoning beyond a department of mathematics, recognizing and celebrating the need and utility of quantitative approaches in thinking and reasoning.

— David Brakke

Peter Ewell of the National Center for Higher Education Management Systems (NCHEMS) argues that the dynamics—"and therefore the levers"—for change are quite different in higher education than in the K–12 sector. In higher education, he suggests, the QL movement must be led largely from outside mathematics, relying on practitioners in client disciplines who really understand and practice QL. Needs will differ in the sciences and engineering (where calculus-based methods dominate) and the social sciences and business (where statistics and applied numeracy are paramount).

One of the important characteristics of quantitative literacy is that it must be situated in the contexts and disciplines where it is used (including mathematics!). This poses a new kind of policy challenge, since one cannot rely on the old paradigm of departments and specialists as a force for change. QL is not a discipline but a way of thinking that influences what, how, and why students learn what they do. It requires creativity in assessment, since neither course grades nor test scores provide a reliable surrogate. However, QL also creates opportunities to enlist support from many fields. "It is encouraging that so many faculty and so many courses can contribute to the acquisition of QL," comments former AAHE president Russell Edgerton, but it is also discouraging because "our internal mechanisms for quality assurance (e.g., curriculum review committees) are overwhelmingly focused on what subjects are taught, not on whether subjects are taught in ways that help students acquire core abilities."

As campus discussions unfold concerning the role of QL in college requirements, prerequisites, and curricula, several issues inevitably emerge. One concerns related goals such as analytical or critical thinking. On some campuses, critical thinking has become its own program with identifiable faculty and courses. Steering QL in the direction of critical or analytical thinking could help ameliorate what some perceive to be the restrictive and sometimes negative connotations of the word *quantitative*, but moving too far in this direction risks losing the unique strength of quantitative methods which, at their best, discipline judgments and decisions. The quantitative dimension is a necessary component of tough-minded critical thinking.

Another worry—especially strong among mathematicians—is that QL is (or may become) merely a remedial enterprise. This view is shared by idealists who believe that secondary schools should provide students with a complete general education and by elitists who do not wish to see university education demeaned by what they consider to be lower learning. Both these views, however, rest on a narrow and simplistic view of quantitative literacy. QL functions at many levels, and no matter how much is accomplished in secondary education there will be much to pursue at the tertiary level that is every bit as sophisticated and subtle as other subjects that students study in higher education. Most opportunities to employ mathematical concepts are not available to students at the same time that mathematical skills are first learned: only through repeated use in increasingly complex circumstances can these mathematical skills become QL habits of mind.

Third is the issue of leadership: Is QL centered somewhere—perhaps in the mathematics department or a general education program—or is it diffused throughout the curriculum as an embedded aspect of a great number of courses? With few exceptions, most faculty who take up this issue advocate that QL should follow the lead of writing as "across the curriculum" or "in the disciplines" (see sidebar). The interdisciplinary and contextual nature of QL cries

Embedding QL Across the Curriculum

Quantitative literacy cannot be taught by mathematics teachers alone, not because of deficiencies in teaching but because quantitative material must be pervasive in all areas of students' education. Quantitative literacy is not simply a matter of knowing how to do the mathematics, but also requires the ability to wed mathematics to context. This ability is learned from seeing and using mathematics regularly in contexts outside the mathematics classroom: in daily life, in chemistry class, in the business world. Thus, quantitative literacy cannot be regarded as the sole responsibility of … mathematics teachers.

A persuasive argument can be made that the skills component of quantitative literacy is essentially precollege in nature. What, this argument goes, beyond the topics of precollege education (graphs, algebra, geometry, logic, probability, and statistics) is foundational to quantitative literacy for everyday life? Looking at the curriculum as a list of topics, however, misses an important point: quantitative literacy is not something that one either knows or does not know. It is hard to argue that precollege education in writing fails to cover the basics of grammar, composition, and voice, for example. Yet it is widely accepted that writing is a skill that improves with practice in a wide variety of settings at the college level.

Quantitative literacy at the college level also requires an across-the-curriculum approach, providing a wide variety of opportunities for practice. The challenges to incorporating quantitative literacy across the curriculum are many, including "math anxiety" on the part of both faculty and students, lack of administrative understanding and support, and competing pressures for various other literacy requirements. Nonetheless, the ability to adapt mathematical ideas to new contexts is a key component of quantitative literacy. The everyday life component of quantitative literacy argues forcefully for engagement of faculty across the curriculum. Quantitative literacy thus must be the responsibility of teachers in all disciplines, and cannot be isolated in mathematics departments.

— Randall M. Richardson & William G. McCallum

out for a crosscutting approach. For some, this conclusion is supported also by a fear that QL in the hands of mathematicians alone would devolve into the kind of other-worldly courses that created the need for new approaches to numeracy education in the first place.

Although many advocates of QL express frustration at the apparent indifference of mathematicians and mathematics departments to the QL needs of students, others firmly recommend that leadership for QL must emanate from mathematics departments. "If we make distinctions that can be translated as 'quantitative literacy is not mathematics,'" argues mathematician William Haver, "we run the risk of giving ammunition to those who oppose reforming the mathematics curriculum and instruction in ways encouraged by quantitative literacy advocates. … If quantitative literacy is viewed as not a central part of mathematics, then it will be much more difficult to direct the energy and resources of mathematicians and mathematical sciences departments toward this important effort."

From this perspective, QL might be seen as a Trojan horse for reform of undergraduate mathematics. Even those who do not go that far seem to agree that mathematics departments should take leadership for QL education, but not total ownership. If QL remains the responsibility solely of mathematics departments—especially if it is caged into a single course such as "Math for Liberal Arts"—students will continue to see QL as something that happens only in the mathematics classroom. If QL is ignored by the mathematics department, or assigned low priority, it will never gain traction on campus as an important objective, and as we have previously observed, will also be ignored by secondary schools. Clearly the best approach is for mathematicians to lead a QL coalition of faculty from across the institution.

Leadership from mathematicians is important for another reason as well: to develop faculty capacity for teaching mathematics in context. Surprisingly many faculty who may use quantitative techniques in their own field have never reflected on the methods they use; some even harbor real anxieties about mathematics, although they can use learned techniques in their own field. Campus-based workshops involving mathematicians, statisticians, and subject experts from other departments can do a great deal to promote faculty development in quantitative literacy: workshops not only enhance faculty ability to teach for QL, but also facilitate campus discussion of expectations, programs, and requirements.

Finally, it is important that QL be implemented across the curriculum because faculty in all disciplines are easily tempted to believe that since the underpinnings of quantitative literacy appear to be topics from middle school mathematics—percentages, ratios, rates—college faculty (and even high school faculty) are not responsible for QL. This argument might be plausible if QL were a skill such as long division or factoring. But it is not: it is a literacy whose functionality grows throughout life. As faculty from different disciplines come to appreciate the subtleties of QL and as they see it used in a variety of contexts, they will increasingly appreciate its power as a college-level literacy. Although many of the mathematical foundations of QL are laid in middle school, numeracy can only be developed by a continued, coordinated effort throughout secondary school and college analogous in intensity and goals to the commitment colleges now make to improve their students' skills as writers.

Educational Goals for Quantitative Literacy

> Quantitative literacy describes a habit of mind rather than a set of topics or a list of skills. ... [QL] is not about how much mathematics a person knows, but about how well it can be used.
>
> — Deborah Hughes Hallett

> An important part of [QL] is using, doing, and recognizing mathematics in a variety of situations.
>
> —Jan De Lange

> The QL skills of interpreting and discussing data and then presenting information in a coherent manner are absolutely essential if our young people are going to be successful in this new world of technology.
>
> —Charlotte Frank

As literacy implies an ability to use words to comprehend and express ideas, numeracy implies a similar capacity to employ numbers and grasp the concepts they represent. Both involve interpretation, understanding, and the power of language. Both mature as students move from one tier of education to the next. Both are essential for democratic life in the information age. And both must be recognized as equal but parallel goals of undergraduate education.

As with reading and writing, the third R can be explored on many different levels. At the lowest level, numeracy is exemplified by common experiences such as determining how much paint or carpet to purchase or deciding which cell phone plan is best. Yet even these contexts are usually much more complex in practice than one might suspect from their simplistic caricatures in school texts or standardized tests. There is a very large difference in sophistication, for example, between calculating the carpet needed for a simple rectangular room (a common test question) and calculating the carpet required for adjoining L-shaped rooms when the carpet has both horizontal and vertical repeat patterns. Both require just simple arithmetic, yet the latter—the more realistic case—involves careful reasoning with scale drawings in order to determine just what calculations are appropriate.

The difference in cognitive complexity between carpeting a rectangular room and measuring carpet for realistic floor plans corresponds in curricular terms to the difference between primary and secondary education. Although one rarely sees problems such as realistic carpet measurement among the stated goals of K–12 mathematics education—factoring polynomials and solving quadratic equations are much more common expectations—the problem-solving demands are approximately the same in both. Laying carpet and solving quadratics are both algorithmic: they require only careful application of a series of steps that can themselves be practiced separately and which, when combined in the proper order, will yield a solution to the problem at hand.

Important as these algorithmic abilities may be for daily life, quantitative literacy is not really about them but about challenging college-level settings in which quantitative analysis is intertwined with political, scientific, historical, or artistic contexts. Here QL adds a crucial dimension of rigor and thoughtfulness to many of the issues commonly addressed in undergraduate education.[15]

Like writing, QL has many faces, each suited to different purposes. Unlike biology or mathematics, QL is not a discipline but a literacy, not a set of skills but a habit of mind. As physics and finance depend on mathematical tools, so too does quantitative literacy. But as these disciplines differ from mathematics, so too does quantitative literacy.

Examples of important issues that can benefit from quantitative understanding are not hard to find. The sidebar, "Making Representative Democracy Work," outlines the roots of quantitative thinking in the early years of the United States. Contemporary examples appear regularly in news headlines:

- *Political polling*. How can polls be so accurate? Why do they sometimes fail?
- *Clinical trials*: Why is a randomized double blind study the most reliable?
- *Tax policy:* Can lower tax rates yield greater tax revenue?
- *Vaccination strategy*: Ethics of individual vs. societal risks (e.g., smallpox).
- *Investment strategies:* The logic of diversification vs. the psychology of risk.
- *Improving education*: What data are required to infer causation from correlation?
- *Fighting terrorism*: Balancing lives vs. dollars and other incommensurate comparisons.
- *Cancer screening*: Dealing with false positives when disease incidence is low.
- *Building roads*: Why the "tragedy of the commons" often leads to slower traffic.
- *Judging bias*: Dealing with Simpson's paradox in disaggregated data.
- *Clinical trials*: Ethics of using placebos for seriously ill patients.

Except for a very few items (where a graph can substitute for algebraic analysis), thorough knowledge of these examples does not require advanced mathematics. The mathematics undergirding these examples actually repre-

Making Representative Democracy Work

Representative democracy originated in a numerical conception of the social order, under the US Constitution. That same document ordained that government should "promote the general welfare and secure the blessings of liberty," a mandate that around 1820 was increasingly answered with a turn toward "authentic facts" and statistics. Statistics soon became compressed into quantitative facts, an efficient and authoritative form of information that everyone assumed would help public-spirited legislators govern more wisely. Schools, both public and private, correspondingly stepped up arithmetic instruction for youth, bringing a greatly simplified subject to all school-attending children, making it possible for them to participate with competency both in the new market economy and in the civic pride that resulted from the early focus on quantitative boasting.

As basic numeracy skills spread, so did the domain of number in civic life. The unsophisticated empiricism of early statistical history yielded to a more complex political terrain where numbers were enlisted in service of political debates and strategizing. At mid-century, the level of quantitative mastery required to keep up with debates based on numbers was still within the reach of anyone schooled to long division and percentage calculations. At a deeper level, the quantitative savvy required to challenge numbers (for bias, for errors in measurement and counting, for incorrect comparison of figures, for selective use of numbers) was not well developed, either in the producers or consumers of numbers. Choices about what to count and what not to count might be made naively, or purposefully and politically, as in the decision not to collect comparable demographic data on blacks and whites in the census.

— Patricia Cline Cohen

sents a rather complete alternative to traditional secondary school mathematics: linear and exponential functions abound, as do graphs, probability, elementary statistics, and exploratory data analysis. Although some traditional topics are hardly present (e.g., trigonometry and binomial theorem), others such as contingency tables and false positives are common in these examples—yet rarely covered in traditional K–12 curricula. A student whose secondary school mathematics included the tools necessary to deal with these QL issues (but without mastery of trigonometry) may be unprepared for calculus but well prepared for the world.

Many Europeans take a broader view of mathematics, arguing that in education the difference between QL and mathematics should be minimal (e.g., de Lange, Niss). Indeed, what the OECD Program for International Student Assessment (PISA) calls mathematical literacy is very similar to what we are calling quantitative literacy, namely, the "capacity to identify, to understand, and to engage in mathematics and to make well-founded judgments about the role that mathematics plays …[for] life as a constructive, concerned and reflective citizen."[16] Through this new assessment, the international community is saying that to serve a broader range of students, mathematics education must become more like QL.

It is easy to identify several key features of quantitative literacy. *Engagement with the world* is the most important. Whereas mathematics can thrive in an abstract realm free from worldly contexts, QL is anchored in specific contexts, often presented through "thick descriptions" with rich and sometimes confusing detail. (Contrast this with the thin gruel of so-called "application problems" common in mathematics textbooks.) Often the contexts of QL are personal or political, involving questions of values and preferences. What is the fairest way to run an election? Are the risks of mass vaccination worth the potential benefits? For students, contexts create meaning. Yet all too often, mathematics students fail to see the relevance of their studies. "If we try to teach students the right competencies but use contexts that are wrong for the students," writes Dutch mathematician Jan de Lange, "we are creating a problem, not solving it."

A second cornerstone of QL is the ability to apply quantitative ideas in *unfamiliar contexts*. "Employees must be prepared to apply quantitative principles in unforeseeable contexts," writes Linda Rosen, former vice president of the National Alliance of Business. This is very different from most students' experience in mathematics courses where the vast majority of problems are of types they have seen before and practiced often. "An essential component of QL is the ability to adapt a quantitative argument from a familiar to an unfamiliar context," observes University of Arizona mathematician Deborah Hughes Hallett. Dealing with the unpredictable requires "a mind searching for patterns rather than following instructions. A quantitatively literate person needs to know some mathematics, but literacy is not defined by the mathematics known."

Third, quantitative literacy insists on flexible understanding that adapts readily to new circumstances. For mathematicians and mathematics teachers, conceptual understanding is the gold standard of mathematical competence. Quantitative literacy requires even more: a flexible understanding that empowers sound judgment even in the absence of sufficient information or in the face of inconsistent evidence. A civil servant timing meter ramps on freeways, a business owner investing in new equipment, and a patient deciding on a treatment for prostate cancer all must make quantitatively based decisions without the kind of full information that they have come to expect from their experience of mathematics in school. The flexibility demanded by QL is not unlike what the Mathematics Learning Study Committee of the National Research Council calls "adaptive reasoning."[17]

Notwithstanding widespread agreement that QL requires engagement with the world, flexible understanding, and capacity to deal with unforeseeable contexts, there still remains some disquiet about the term itself. The facility we want students to acquire is not just about quantities, nor is it as passive as "literacy" may suggest. Reasoning, argument, and insight are as essential to QL as are numbers; so too are notions of space, chance, and data. Like the international team that is designing PISA, many prefer the term "mathematical literacy" since it does not suggest a restriction to the quantitative aspects of mathematics, but invites a role for mathematics in its broadest sense.

The word "context" also causes confusion. Since context is a catalyst for promoting learning, good teachers always provide context. So in some sense, teaching "in context" is redundant: it just means "teaching well." But contexts can be quite varied and only a few may appeal to a particular student at a particular time. So although learning in context is usually easier than learning without context—at least if the context is relevant to the student—teaching in context is not at all easy. As Russell Edgerton observed, the more context dominates content, the more the instructional emphasis can shift from teaching to learning. All too often, unfortunately, feeble attempts at context produce minds that are "not in gear:"

> A quantitatively literate person must be able to think mathematically in context. This requires a dual duty, marrying the mathematical meaning of symbols and operations to their contextual meaning, and thinking simultaneously about both. It is considerably more difficult than the ability to perform the underlying mathematical operations, stripped of their contextual meaning. Nor is it sufficient simply to clothe the mathematics in a superficial layer of contextual meaning. The mathematics must be engaged with the context and providing power, not an engine idling in neutral. Too many attempts at teaching mathematics in context amount to little more than teaching students to sit in a car with the engine on, but not in gear. The "everyday life" test provides a measure of engagement; everyday life that moves forward must have an engine that is in gear.
>
> —Randall Richardson and William McCallum

As students move from school to college and beyond, so does their potential need for quantitative literacy. As students move through education and on into adult life, issues they face in finance, politics, and health increase in subtlety and sophistication, yet often rely only on mathematical techniques of ratios, averages, and straight line graphs that were first learned in the middle grades. It is this increased sophistication built on a foundation of relatively elementary mathematics that is a hallmark of quantitative literacy.

Much the same is true for writing: although the principles of grammar and organization are more or less established during school-level instruction, students' sophistication in writing continues to grow through college and well beyond. Experienced educators know approximately what to expect of student writers at different grade levels. This is not yet true, however, for quantitative literacy because in its present form QL is largely new—both as a requirement of engaged living and as a goal of education.

So one important task required to establish QL as an effective goal for undergraduate education is to define what QL means at different educational levels. Rita Colwell, Director of the National Science Foundation, argues that the nation needs "flexible goals" for QL based on standards that are appropriate for different audiences. "Most importantly," she notes, "we must understand that literacy has levels." For QL to take root in higher education, we need to understand and express how college-level QL differs from high school level QL, and how both differ from the middle school canon of simple percentage and interest problems.

Importance of Quantitative Literacy

> In our work-based society, failure to give people the knowledge and skills they need to get and keep good jobs can have disastrous personal consequences. —Anthony Carnevale

> Careers and work are not the only economic reasons for taking mathematics. Nonetheless, the primary force behind the nation's emphasis on mathematics is economic. — Arnold Packer

> Many individuals are not prepared to function effectively in the workplace today because they lack versatility and flexibility on how they approach and deal with obstacles. —Linda Rosen

The case for improved quantitative literacy is often argued in terms of economic competitiveness, both for individuals and for society. While this case is legitimate and strong, a more general issue underlies our concern in this report: the increasing importance of quantitative data for each person's quality of life and for our collective well-being. We see this not just in terms of jobs and workforce issues, but also for everyday issues of personal welfare, social decision-making, and the functioning of democratic society.

Personal welfare includes issues such as health, safety, taxes, budgets, credit, and financial planning. Health and safety are obviously important, as we have seen in the reduced effectiveness of antibiotics, the spread of viruses to new regions of the world, and recently, the threat of bioterrorism. So too are issues of family finance, especially as consumer choices expand for household utilities, health insurance, and retirement plans. Decisions about health and finance virtually always involve quantitative reasoning. They require average citizens to read meaning into numbers—to assess risks, to make and follow budgets, and to understand how planning projections depend on assumptions. As literacy empowered peasants of earlier centuries to exercise personal choice and gain some measure of autonomy, so in our age numeracy offers average citizens the opportunity to make intelligent choices and exercise some degree of control over what otherwise would appear to be a totally mysterious domain of numbers.

Similarly, effective democracy in this age and in this country requires an alert and skeptical citizenry capable of dissecting political arguments and evaluating economic discourse. Increasingly, quantitative skills such as reading graphs, understanding models, and estimating rates of growth are important for active citizens who seek to influence their government and protect their rights. So too are issues of representation and voting, which rest on mathematical principles that are often (and unwisely) taken for granted. Quantitative literacy is an essential element in many duties of citizens: evaluating allocation of public resources, understanding media information, serving on juries, participating in community organizations, and electing public leaders. As ETS vice president Anthony Carnevale argues, its absence can be devastating:

> The wall of ignorance between those who are mathematically and scientifically literate and those who aren't can threaten democratic cultures. The scientifically and mathematically illiterate are outsiders in a society where effective participation in public dialogue presumes a grasp of basic science and statistics. Their refuge is a deep mistrust of technocratic elites that often leads to passive withdrawal from public life or an aggressive and active opposition to change.... Citizens who are resigned to being cogs in some incomprehensible machine are not what the founders of the American republic had in mind.

Often headlines aimed at innumerate readers create a distorted sense of risk that seems to control public mood and political responses to various crises, real or imagined: anthrax, snipers, shark bites, irradiated food, or cell phone radiation. Does the public understand how these risks compare with gunshots, automobile accidents, tornadoes, or smoking? Quantitatively alert journalists should provide accurate information about relative risks, just as they now strive to provide accurate information about their sources' professional positions and titles. The latter are intended to warrant credibility by inference from an individual's title; the former (statements of relative risk) would enable readers to make judgments for themselves—certainly a civic virtue worth cultivating.

Public policy is increasingly influenced by reports from a growing infrastructure of "think tanks," many representing special political interests. Their primary purpose is to employ quantitative data to influence public policy across a wide spectrum of domains (e.g., politics, health care, economic policy). This recent extension of quantitative methods in the public sector elevates significantly the level of quantitative literacy a citizen needs in order to participate in the political process.[18] The growth of these agenda-driven research centers, and of their influence, demonstrates a need for QL that is drawn from the environment in which we now live, rather than only from poor test scores documenting the "Johnny can't add" critique of curricula designed for environments in which we used to live.

Limnologist David Brakke, Dean of Science and Mathematics at James Madison University, makes this point in the context of how science influences public policy:

> If science is to inform policy, we must ask the right questions, collect appropriate data, and conduct analysis in a decision-making framework. Such a process would naturally involve design, sampling, error, estimation, and uncertainty. We might consider rates, variability, predictability, scales, and limits. We might need to evaluate actual and perceived risk and find ways to manage those risks. Ultimately, we must communicate the results. How do we best communicate the results of an assessment of risk, especially when a public comfortable with probabilities in weather forecasts is expecting science to have definite, certain answers? Informed decision-making in a world full of data requires quantitative reasoning.

By tracing the historical balance of quantitative power between government bureaucracies and average citizens, historian Patricia Cline Cohen foresees an alarming shift in favor of bureaucracies:

> The post-Civil War era finally brought a full melding of statistical data with the functioning of representative government. A century after the first census of 1790, no one suggested any longer that an expanded census would alarm the people or merely gratify idle curiosity. The government had accepted an on-going obligation to monitor the vital signs of the nation's health, wealth, and happiness. The census bureau was at last turned into a permanent federal agency, ... [However], this growing sophistication of government statistical surveillance was not matched by a corresponding improvement in quantitative literacy on the part of the public. Unlike the early 19th century, when a public enthusiasm for numbers and arithmetic developed along with a statistical approach to civic life, in the early 20th century the producers of statistics quickly outstripped most consumers' abilities to comprehend. The number crunchers developed more complex formulations, while the arithmetic curriculum stagnated.

At the beginning of the 21st century, as government and think tank number crunchers use yet more sophisticated methods, so the QL demands of alert citizens have moved well beyond mere arithmetic. Sadly, public education has not kept pace.

Data on wages and careers deliver the same message. According to ETS's Anthony Carnevale, quantitative and mathematical ability is the best predictor of the growing wage advantages from increased postsecondary educational attainment. Even among students who do not intend to pursue programs of study that require advanced quantitative skills, success in high school mathematics strongly influences access to selective colleges. At the same time, the role of college-level quantitative skills in allocating economic opportunity is of growing significance. And as Carnevale notes, this linkage is particularly significant in the United States since here it is poorly educated individuals, rather than employers or governments, who pay the price of educational inequality. "This is quite different from continental European labor markets, which have inherent incentives to educate and train all workers in the hope that their productivity will justify earnings and benefits guaranteed by the welfare state."

The leveraging effect of quantitative literacy can be seen in many industries that have benefited from the dramatic recent advances in mathematical, statistical, and computational methodologies. These include, most visibly, finance (where the valuation of risk[19] has made possible both boom and bust on Wall Street), law (dealing with such issues as disparate impact, legislative redistricting, and statistical evidence[20]) and cinema (virtual actors, computer-generated dinosaurs[21]). A specific example cited by David Brakke is the "New Biology":

> In the world of biology we have become data-rich, with new horizons requiring new sets of skills. We have genomics, nanotechnology, biomaterials, DNA computing, neuroscience and cognition, and environmental science (or biocomplexity) in all dimensions and scales. This "New Biology" is multi-dimensional and collaborative, multi-disciplinary, information-driven and education-oriented. Modeling, managing with information, recognizing patterns in vast amounts of data, all require sophisticated mathematical and computational skills. Mathematics, statistics, and computational science become essential elements of biology, determining anew what quantitative skills are needed.

Details of the increased significance of quantitative tools for biology and medicine are enumerated in a new report from the National Research Council entitled *Bio2010*.[22] For example, new mathematical ideas are helping us understand what determines whether an epidemic waxes or wanes—a question, interestingly, that launched one of the earliest American efforts at quantitative analysis during outbreaks of small pox in the eighteenth century.[23] Quantitative reasoning also helps explain how leopards get their spots, how chance events can markedly alter gene frequencies in small populations, and how lengthy strands of DNA can be packed into a cell nucleus. College graduates who are quantitatively literate will be comfortable working in the new biology, or the new policy arena, or in any field that depends on data for understanding and creativity.

Paradoxically, during the last half-century as access to postsecondary education has increased, the gap between the best and the average graduate has widened. In mathematics we have the best graduate schools in the world but among the least capable high school graduates. In higher education, nearly three out of every four mathematics enrollments are in courses that repeat the mathematics taught in secondary school, yet at the same time the most rapidly growing area of secondary school education are courses intended to provide college credit.[24] In comparison with other high income countries, the inequality between the best and poorest performers in the United States is very large.[25] Although the U.S. has proven itself capable of producing experts in a wide variety of fields, we seem not to have learned how to prepare graduates for the QL-demands of 21st century jobs or to nurture a literate citizenry needed to sustain a robust democracy.

We have many grounds for concern. Estimates that 60% of all new jobs in the early 21st century will require skills possessed by only 20% of the current workforce is one such cause for alarm.[26] In comparison with countries of similar per-capita income, U.S. adult literacy scores are about average. However, U.S. adults over age 35 score well above the international average, and those under 35 score well below average.[27] This does not bode well for the future.

On average, U.S. adults have acquired more years of formal schooling than their international counterparts, but do not match their literacy performance at any level of schooling. More worrisome, the literacy of U.S. adults is more varied than in virtually any other country: U.S. adults are among the best internationally at the top of the literacy scale, but among the worst at the bottom.[28] The most recent International Adult Literacy Survey (IALS) indicates that the literacy rate in the United States is the second lowest of 18 European/North American countries. Moreover, 58% of U.S. high school graduates—the highest in the study—perform on literacy tasks below the minimum level needed to cope adequately with the complex demands of everyday life. (The median for the other countries is about 20%.)[29]

Quantitative literacy fares not much better in higher education. Only rarely is QL thoughtfully integrated into general education, in part because QL is not a central priority for any discipline and general education is rarely a central priority for mathematics departments. Often, general education is mistakenly equated with elementary (so-called "service") courses in mathematics, statistics, or other quantitative disciplines. Typically, students meet a quantitative requirement either by passing a test of basic skills or by taking a course in mathematics or statistics designed for a pre-professional purpose.

The mathematics courses most commonly used by students to satisfy a general education requirement—college algebra or calculus—have almost nothing to do with anything that will affect their future quality of life or their ability to contribute to democratic society. Even the typical statistics course, although better, leaves most students with at best a mechanical repertoire of hypothesis tests that are of little value to average citizens. What students most need these courses rarely provide, namely, extensive practice in developing the numeracy skills of data analysis and evidence-based criticism in the context of various disciplines. As a consequence, there is very little evidence to suggest that students leave college much more numerate than when they enter.

The result is widespread adult innumeracy, often reinforced by thoughtless media reports of quantitative information. A common example is the misunderstanding of the significance of SAT averages, even by professionals with graduate degrees. Solemn news articles describe variations in these averages among states, politicians argue for or against tax increases based on changes in these averages, and parents decide where to live and how much to pay for a home—all without thoughtful interpretation of the enormous variation in the size and characteristics of the population of test-takers from state to state or year to year.[30] Judgments about data can differ radically and require a good deal of sophistication to evaluate, especially concerning the future of Social Security, the differential and future effects of tax cuts, the flow of immigration into the country, the rising or falling of student test scores, and the gyrations of the stock market as summarized in a few one-number indices. "Both politicians and voters may be in over their heads when it comes to evaluating different projections," notes Patricia Cohen. "The danger is that we may not realize we are in over our heads."

For similar reasons, public health officials worry that patients' innumeracy can lead to misinterpretation of medical advice or health-related news, while patients who follow news reports about dosage errors in hospitals worry about the innumeracy of medical personnel. Judges wonder if average citizens can perform the minimal civic duties of jurors in weighing evidence that, increasingly, is presented in quantitative and scientific terms.[31] Studies of the labor market find that well-paying jobs in which there are the most significant labor shortages tend to be those requiring the strongest QL skills.[32, 33] And scores from national assessments show surprising weaknesses in young adults' abilities to carry out even basic quantitative tasks of daily life.[34] Examples abound where innumeracy handicaps individuals—and thus society—in health, citizenship, and careers.

Quantitative Literacy and Mathematics

> The historical reasons for situating the learning of QL skills in mathematics study have not lost their relevance. —Hyman Bass

> From school to college, mathematics follows an isolated trajectory of increasing difficulty and abstraction whose apparent purpose is to select and prepare the best mathematics students for ... mathematically intensive fields. —Anthony Carnevale

> Should a commitment to quantitative literacy replace, supplement, or transform the mathematics curriculum? This seems to be one of the basic questions. —Andrea Leskes

For better or worse, the notion of quantitative literacy is associated in the minds of most educators, politicians, and parents with mathematics. Occasionally, especially if prompted by language in state standards, an association is also made with statistics, a subject that is slowly becoming accepted as part of school mathematics. Colleges often allow students to use courses in either mathematics or statistics to fulfill the QL part of a general education requirement. Thus on both sides of the school-college boundary, policy related to quantitative literacy is inextricably intertwined with mathematics and statistics.

As the public correctly senses, many of the skills necessary for numeracy can be found in various mathematics courses. However, although some elementary mathematics is required for QL, advanced mathematics is not at all the same as QL. Whereas QL is primarily about reasoning with data, mathematics is about patterns, numbers, and space. Although mathematics is a fundamental subject of real value to all students, most mathematics courses are ill suited to achieving quantitative literacy. "The ability to perform some of the basic operations of mathematics is necessary for quantitative literacy," write Randall Richardson and William McCallum of the University of Arizona, "but even the ability to perform many of them is not sufficient. Anyone who has taught mathematics, or who has taught a subject requiring the mathematics that students have learned in previous courses, is aware of this fact. Many students are technically capable but unable to make reasonable decisions about which techniques to apply and how to apply them. Mathematicians and their colleagues in other departments share frustration at the fragility of what students learn in their mathematics classes."

The power of mathematics lies in its generality and abstraction, in its ability to rise above specifics. Quantitative literacy, on the other hand, is anchored in real world data. One consequence, according to astronomer and astronaut George ("Pinky") Nelson, former director of AAAS's Project 2061, is that "current mathematics classes abound in inappropriate, inconsistent, or unrealistic situations and data.... QL-type applications are rare." Grant Wiggins makes a similar point about the isolation of classroom mathematics from things that matter:

> Too many teachers of mathematics fail to offer students a clear view of what mathematics is and why it matters intellectually. Is it any accident that student performance on tests is so poor and that so few people take upper-level mathematics courses? ... Without anchoring mathematics on a foundation of fascinating issues and "big ideas," there is no intellectual rationale or clear goals for the student.

Further, mathematics courses have traditionally been aimed primarily at preparation for future courses, not for immediate use. The dominant K–12 mathematics curriculum is a hurried sequence of courses whose primary goal is to prepare students for calculus. Efficiency of the path to calculus and advanced mathematics has led to rigid linearity of the "GATC" sequence (geometry, algebra, trigonometry, and calculus) in order to sort students effec-

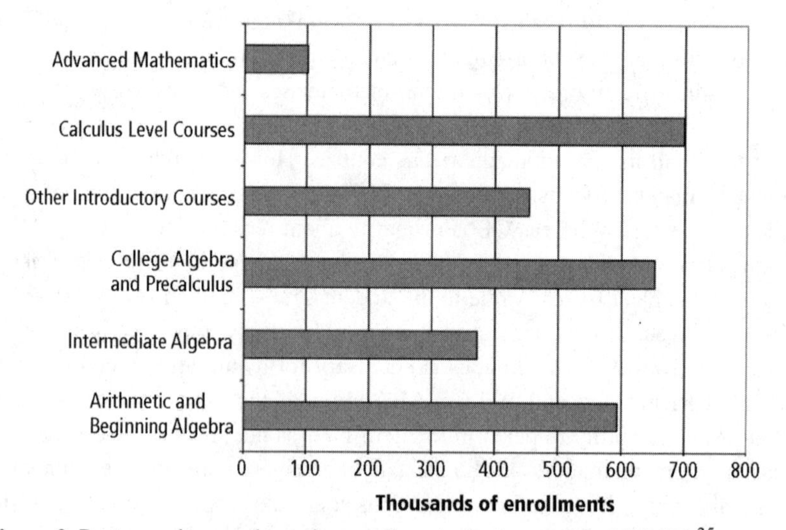

Figure 3. Postsecondary mathematics enrollments (in thousands), fall 2000.[35]

tively. "No other discipline, save perhaps foreign language, exhibits such linearity," notes University of Arkansas mathematician Bernard Madison. "But foreign language education is built on using the language, whereas students' use of mathematics is usually far in the future. Most students in the GATC sequence never get to any authentic uses for what they learn." Fewer than one-fourth of students ever take a calculus-level mathematics course and only about 1 in 30 enrolls in a mathematics course beyond the calculus level.

As measured by enrollments, this GATC sequence is by far the largest part of both high school and college mathematics (see Figure 3). Its rigid linearity leaves little time for teachers to teach mathematics in context or to help students develop the habits of mind necessary to interpret real life situations in quantitative terms. College mathematics courses, Madison argues, must have independent value and not only be "routes to somewhere else." (A more detailed discussion of the GATC sequence and its implications for quantitative literacy can be found in the "insider/outsider" dialogue on college mathematics in Part II.)

The consequences of thoughtlessly trying to move all students through a mathematics curriculum from which fewer than one in ten will emerge into

What is Mathematics?

How is mathematical literacy related to mathematical knowledge and skills? Evidently, that depends on what we mean by mathematics. If we define mathematics in a restrictive way, as a pure, theoretical scientific discipline—whether perceived as a unified, structurally defined discipline, or as a compound consisting of a number of sub-disciplines such as algebra, geometry, analysis, topology, probability, etc.—then it is quite clear that mathematical literacy cannot be reduced to mathematical knowledge and skills. Such knowledge and skills are necessary prerequisites to mathematical literacy but they are not sufficient.

However, this is not the only way to define mathematics. We may adopt a broader—partly sociological, partly epistemological—perspective and perceive mathematics as a field possessing a five-fold-nature: as a pure, fundamental science; as an applied science; as a system of tools for societal and technological practice ("cultural techniques"); as an educational subject; and as a field of aesthetics. Here, being a pure, fundamental science is just one of five "natures" of mathematics. If this is how we see mathematics, then the mastery of mathematics goes far beyond the ability to operate within the theoretical edifice of purely mathematical topics. And then, I submit, mathematical literacy is more or less the same as the mastery of mathematics. By no means, however, does this imply that mathematical literacy can or should be cultivated only in classrooms with the label "mathematics" on their doors. There are hosts of other important sources and platforms for the fostering of mathematical literacy, including other subjects in schools and universities.

— Mogens Niss

courses that use calculus are devastating in human, educational, and social terms. Large numbers of students, including many of society's future leaders, end their study of mathematics with a loathing that approaches phobia.[36] Colleges struggle under the burden of teaching a majority of their students mathematics that is ordinarily offered in high school. For lack of confidence in students' quantitative skills, science faculty often demathematicize their courses, thus diminishing students' opportunities to reinforce skills through application and undermining the argument that quantitative skills are necessary for future success. Too many graduates of this system are unable to comprehend data-based claims about contentious issues in public policy, much less understand risks of medical choices, retirement investments, or even cell phone plans.

Moreover, many adults' last memory of mathematics "still stings many years later," observes Deborah Hughes Hallett. "Whether their last course was in school or college, some remember a teacher whom they perceived as not caring. Some blame themselves for not being able to understand. Some remember a course whose purpose they did not understand and which they perceived as having no relevance. Many remember a jungle of symbols with little meaning." Recognizing the many challenges facing undergraduate mathematics, The Mathematical Association of America recently conducted a comprehensive study of the entire undergraduate program; details and web references are in a sidebar.

Whereas traditional mathematics courses may be too narrow and hurried for effective QL education, courses taught in the name of quantitative literacy are often seen as a lower track for those who cannot succeed in "real" mathematics. This perception—of quantitative literacy as thin mathematics for weak students—does a great injustice to both the sophistication of the problems that depend on quantitative literacy and the educational needs of students. QL is not about "basic skills" or watered-down mathematics, but about a demanding college-level learning expectation that raises significant challenges for the undergraduate curriculum. Experience with struggling students shows convincingly that QL tasks are no less challenging than those encountered in a typical calculus course.

Two factors especially impede students' development of quantitative literacy in traditional courses: the difficulty of transferring formal knowledge to everyday contexts and the gap in sophistication between important contexts and necessary skills. Since learning is inherently associative, it is not surprising that students have difficulty applying a skill that they associate with one setting to something entirely new. This problem of transference is notorious-

Undergraduate Mathematics in the 21st Century

Coordinated by its Committee on the Undergraduate Program in Mathematics (CUPM), the Mathematical Association of America (MAA) spent four years conducting a major review of undergraduate mathematics programs ranging from teacher education to quantitative literacy and from service courses for partner disciplines to expectations of students who intend to a pursue mathematically-intensive careers. The CUPM report, to be issued at the end of 2003, builds on earlier special studies concerning quantitative reasoning,[37] the mathematical preparation of teachers,[38] and the mathematical needs of partner disciplines.[39]

The CUPM report urges colleges to build their mathematics programs on a foundation of student needs. Their recommendations focus specifically on students who take mathematics only as part of general education (on most campuses, the overwhelming majority); or as part of quantitative requirements for partner disciplines; or to prepare for careers as K-12 teachers; or to major in mathematics for a variety of possible careers.

Several CUPM subcommittees are preparing to follow-up with specific assignments. Two are of special importance to QL: the Quantitative Learning subcommittee (URL: http://www.valpo.edu/home/faculty/rgillman/ql/) whose special focus is QL, and the subcommittee on Curriculum Renewal Across the First Two Years (CRAFTY) which has been charged with reconciling the curricular dilemmas posed by college algebra, precalculus, and quantitative literacy. The MAA is also expected to set up a Special Interest Group (SIG) in quantitative literacy to encourage networking among its members with interests in QL.

ly difficult when it involves applying abstract skills learned in mathematics courses to real contexts outside a formal setting.[40]

A second impediment is perhaps less widely recognized. Even though the mathematical and statistical tools underlying everyday quantitative situations are often quite elementary, the contextual uses confronting adults in daily life are surprisingly sophisticated, often surpassing the abilities of most high school and many college graduates. To be useful to adults, skills learned early in school need to be applied or reinforced in more sophisticated contexts throughout the educational system. Without such reinforcement, elementary skills will remain associated with elementary uses (e.g., school "word problems") that are not the kinds of numeracy problems that confront adults.

College Algebra

For historical reasons, quantitative literacy is intimately linked with the course that colleges call college algebra or precalculus and that secondary schools call, variously, advanced algebra, precalculus, or algebra and trigonometry. Despite all that one hears in the press about colleges expecting entering students to have completed AP calculus, most students begin their undergraduate mathematics experience with college algebra. It is, de facto, the most common general education course in mathematics—and by almost all accounts, a failure.[41] University of Arkansas mathematician Bernard Madison, a former dean of arts and sciences, summarizes the indictment succinctly:

> The traditional college algebra course is filled with techniques, leaving little time for contextual problems. Students, many of whom have seen this material in prior algebra courses, struggle to master the techniques; three out of four never use these skills, and many of the rest find that they have forgotten the techniques by the time they are needed in later courses. No wonder the course is uninspiring and ineffective. Success rates are very low—often below 50%—and student dissatisfaction is high.
>
> In the interest of efficiency, we have gathered together largely uninspiring algebraic methods and created courses with a singular, dominating goal of preparation for calculus ... Those who do not survive are left on the side of this narrow road with fragmented and often useless methodological skills.

Focused entirely on a wide range of relatively specialized algebraic techniques that students rarely remember beyond the final exam, college algebra neither prepares students well for courses in other quantitative disciplines nor for their civic, employment, or personal needs.

Moreover, college algebra sits in the center of the mathematics curriculum, straddling the boundary between school and college. Since mathematics placement exams are geared to college algebra and its prerequisites, college algebra effectively controls the mathematics placement of entering college students and sends powerful signals to high schools about the mathematics colleges value. In most universities college algebra also serves as a way to support graduate students who are employed to teach the course. It functions, therefore, as a well-institutionalized means of subsidizing graduate programs with undergraduate tuition dollars.

It is difficult to overstate—and there is no way to overlook—the vast differences in educational goals reflected on the one hand by the demands of quantitative literacy and, on the other hand, by the vast enterprise of courses, prerequisites, and requirements for college algebra. Janice Somerville, Executive Director of the National Association of System Heads (NASH), expresses frustration over mathematicians' lack of attentiveness to this issue. "Frankly, I am puzzled by the relative silence about first making clear the place of QL in mathematics departments." Thinking optimistically, mathematician Donald Small of the United States Military Academy sees the entrenched status of college algebra as a potential opportunity for QL:

> I feel strongly that traditional college algebra should be reformed to serve as a base for QL programs … It is a well established course that is widely respected (outside of mathematics departments) and that enrolls more students than any other college mathematics course ... By reforming college algebra we can avoid both the political problem of finding a home for a new course and the practical problems of attracting students and developing faculty support.

These many problems have led to a series of national conferences and planning meetings dealing solely with the problems of college algebra and its relation to precalculus and quantitative literacy.

On some campuses there is a move to establish two versions of college algebra—one for students who need calculus for their proposed majors, the other for students who do not (and who presumably are taking college algebra primarily to fulfill a QL-inspired college graduation requirement.) Freed from the constraint of including precalculus techniques, the alternate course could move in the direction of quantitative literacy, perhaps with a focus on mathematical modeling.[42] Such a course, advocates note, could focus not on precalculus algorithms but on data, technology, and quantitative communication. It could serve as the hub of a campus-wide QL program in much the same way as freshman composition anchors writing-across-the-curriculum programs.

Caution is in order, however. Most problematic is that two versions of college algebra will reinforce the tradition of tracking that has arguably created many of the problems besetting mathematics education in the United States. It also assumes that students in the mainstream algebra track are (or will become) quantitatively literate, or that it does not matter if they do not. Reports from various campuses offer little evidence in support of the former assumption.

Similar efforts have arisen every decade or so, but none have dethroned the techniques-dominated college algebra. Examples include "Universal Mathematics," the first (and long-forgotten) proposal of the MAA's new Committee on the Undergraduate Program in Mathematics (CUPM)[43]; *Finite Mathematics*,[44] a precursor to what is now often called "discrete" mathematics; and *For All Practical Purposes*, an innovative course from COMAP focused on practical applications of basic mathematics.[45] These courses had one thing in common that contributed to their remaining a small elective rather than a major requirement: they were designed specifically to focus on ideas—generally QL-like ideas—rather than techniques. This made them more difficult for teachers to teach and for students to master, and for that reason they thrived only in special niches out of the mainstream of college mathematics.

Several flawed policies have helped college algebra maintain its hegemony over introductory college mathematics. One is the role it plays in supporting graduate teaching assistants who serve as instructors in major universities: only a mechanically focused course could work with such a distracted and internationally diverse cadre of instructors. Since learning to apply mathematics in unfamiliar situations is very hard, both students and teachers are prone to take short cuts—especially when both have other priorities.

"Students clamor to be shown 'the method,'" writes Deborah Hughes Hallett of the University of Arizona, "and teachers often comply—sometimes because it is easier, and sometimes out of a desire to be helpful. Learning 'the method' may be effective in the short run—it may bring higher results on the next exam—but it is disastrous in the long run."

> Most students do not develop skills that are not required of them on exams. Thus if a course simply requires memorization, that is what the students do. ... Any novel problem or context can be made "old" if students are taught a procedure to analyze it. Then students' success depends on memorizing the procedure, rather than on developing their ability to apply the central mathematical idea. ... There is tremendous pressure on US teachers to make unfamiliar contexts familiar, and hence to make problems easy to do by memorized algorithms. Changing this will take a coordinated effort: both school and college teachers will need to be rewarded for breaking out of this mold.

A. Geoffrey Howson of the University of Southampton, former secretary of the International Commission on Mathematical Instruction (ICMI), ascribes the failings of courses like college algebra to the "piece of tape" mentality in mathematics that assumes that the best curriculum for all is a single tape, increasing in abstraction, that is snipped off at different points for students of different aims. This approach, he argues, "forces weaker students to learn algebraic techniques which they will never develop into usable knowledge. Of course such students are no longer being 'denied' the opportunity to learn algebra, but instead are simply forced to learn techniques that might conceivably (but with a fairly low probability) lead to something more useful and valuable." Carnevale's research supports Howson's concern: "Even a casual analysis of the distribution of occupations demonstrates that relatively few of us—less than five per cent—use geometry, algebra, or calculus on the job.

The techniques-dominated college algebra course is not only the largest course in undergraduate mathematics, but for the majority of its students it is the last mathematics course they ever take. One consequence is that students never discover why it matters if they can't "do" algebra. When they leave academia, they find out for certain that it doesn't matter. Algebra may be the language of mathematics, but it isn't a language most adults ever use. QL is a different story from algebra; many examples illustrate its importance throughout life, whereas these rarely exist for algebra. Pragmatically, mathematics matters to adults only as it concerns their children's college admission. Thus the algebra experience breeds skepticism or cynicism (or both) where any argument about the importance of things quantitative is concerned.

Even calculus, the ostensible purpose of the college algebra/precalculus course, meets very few of the objectives of quantitative literacy. Calculus is no doubt one of civilization's great intellectual achievements, and it is a tool of extraordinary power that is indispensable in many scientific fields. But little in the course addresses the core issues of quantitative literacy—data, inference, context—and even many of the very best calculus students stumble when confronted with QL tasks. The truth is, although QL depends on some simple algebra—less than what is found on the ubiquitous spreadsheet, for example—neither college algebra nor calculus is necessary for QL.

The two major reform movements in mathematics education of the last fifteen years—the emergence of standards for school mathematics and the effort to reform the teaching of calculus—have both moved in the direction of adding meaning to students' experience with mathematics by dealing with authentic data and realistic contexts. Separately, many recent reports on undergraduate education have stressed the importance of engaging all students, not just a select few, in the type of higher order thinking that fosters quantitative literacy.[46] Across the K–16 landscape there are countless ideas and enthusiastic faculty ready to address the problem of innumeracy.

Nonetheless, for a variety of historical reasons, it is mathematics and not numeracy that dominates discussion of educational policy. Mathematics has a venerable history whose value is indisputable and timeless, whereas the value of quantitative literacy is primarily a product of the information age. Mathematics' footprint in the curriculum and assessment is deep and broad; quantitative literacy's is barely discernable. As a priority for education, history is on the side of mathematics; the future however, may rest in the hands of the quantitatively literate.

Statistics

As mathematics is the science of patterns, statistics is the science of data. Thus statistics' link to QL is inherently stronger than that of mathematics. The age of computers is an age of data, mostly numerical data. Even non-numerical data is encoded numerically for transmission and transformation by computers. "Reasoning logically and confidently with data is a crucial component of any curriculum for quantitative literacy," write Randall Richardson and William McCallum. Few would make a similar claim for mathematics (not to mention college algebra).

Whereas mathematics has ancient roots, reasoning with data is a distinctly modern enterprise. Even the word "statistics" is relatively new to the English language, arising as the Enlightenment began to recognize the value of data as a tool for progress, especially for advancing the good of the new democratic, citizen-led states (see sidebar).

Statistics has evolved a great deal in the last two centuries, but has not yet achieved a status in education parallel to that which mathematics has held for centuries. Richard Scheaffer, former President of the American Statistical Association, talks about statistics as the "keeper of the scientific method," as the discipline that studies how to understand the world through setting hypotheses, collecting and analyzing data relevant to those hypotheses, and drawing conclusions about the hypotheses from analysis of the data. "Statistics emphasizes context, design of studies, and a stochastic view of the world."

Unfortunately, practical problems of analyzing data often do not fit nicely the assumptions of statistical theory concerning hypothesis testing that students study in Statistics 101. This difficulty leads directly to the parallel subject of data analysis, which can be thought of as the practice of answering real

The Science of the State

The idea for a state-based survey came from a parallel survey project undertaken in Scotland by Sir John Sinclair, the man who came up with the word "statisticks" and defined it to mean "an inquiry into the state of a country, for the purpose of ascertaining the quantum of happiness enjoyed by its inhabitants, and the means of its future improvement."[47] "Statistick" first appeared in an American dictionary in 1803 with an enigmatic definition referring to Sinclair and his "statement of the trade, population, production [of Scotland]...with the food, diseases, and longevity of its inhabitants." By 1806 the word gained a clearer articulation in Noah Webster's first dictionary as "a statement or view of the civil condition of a people."[48]

The word state was embedded in the word in two ways: facts about the state that could be plainly stated. While such facts might take the form of numbers, this was not yet an essential part of the definition for several more decades. By the mid 1820s, just over a dozen books had been published with the word "statistics" or "statistical" in the title, and another 17 not emblazoned with the word still had the character of reference books of authentic facts and numbers. Nearly all these books proclaimed the novelty of their shared project, assembling facts and figures for the aid of statesmen and citizens.

— Patricia Cline Cohen

questions rather than trying to fit these questions into established theories. "An approximate answer to the right question," goes one mantra of data analysis, "is better than an exact answer to the wrong question."

In reality, even data analysis—what most statisticians actually practice—is more than the average person needs to be an informed citizen, intelligent consumer, or skilled worker. What everyone really needs is typically called statistical thinking or statistical literacy, a crucial component of quantitative literacy. Statistical thinking is the backbone of the contemporary emphasis on quality improvement in the work place. At all levels from office assistant to senior manager, employees in a business must have some notion of statistical thinking if a firm is to operate optimally. Statistical thinking can be thought of as viewing life as made up of processes with variation, says Scheaffer. "Once understood, variation can be broken down into that which can be reduced and that which must be managed.... The inherent variation in processes must be considered in order to see if change can be attributed to any cause other than pure chance."

Traditional school and college mathematics curricula—the "GATC" sequence—move students quickly from algebra to calculus and all but ignore statistics and its close cousin, probability. A quiet movement sustained over the last two decades has begun infusing school mathematics with topics from data analysis, statistics, and probability, often under the name "quantitative literacy."[49] Indeed, in many schools, the term "quantitative literacy" has became a synonym for elementary statistics and exploratory data analysis.

What has not happened is the effective integration of these themes with the core mathematics courses to which college admissions and an increasing numbers of high-stakes state tests are anchored. Although many of the new K–12 curricula contain topics from quantitative literacy, because they are unfamiliar and not required for college entrance they are considered by teachers to be a luxury that can be omitted if time is tight (which it always is). Although many college faculty agree that probability and statistics are vitally important for rigorous college study, none appear willing to tamper with the GATC tradition. "This is a failure of leadership," writes Deborah Hughes Hallett. "The result is that although most adults see probabilistic and statistical arguments every day, few have any preparation to make sense of them."

Leadership for QL

The public looks to mathematics as the discipline best suited to address the quantitative needs of students, and indeed mathematics has the greatest continuity of curriculum from primary grades through college. Although mathematics certainly cannot bear the sole burden of quantitative literacy, it is the discipline best suited to play a leadership role. Although QL is not strictly a part of mathematics, the many current conversations about mathematics education—especially the recent emphasis on course and curriculum renewal—have prepared many mathematics teachers to contribute powerfully to the numeracy needs of society.

Impediments of status, turf, and tradition will inevitably surface. Many students and parents, echoing professors, will look on quantitative literacy as an inferior option, a lower track that can absorb students who appear unable to keep up with the pace of traditional "academic" mathematics. Many mathematicians will resist diverting limited time and resources to lead an educational task they perceive as not central to their discipline, while many faculty outside mathematics will be equally uneasy about entrusting mathematicians with leadership of what is primarily a cross-disciplinary endeavor.

Nonetheless, the needs of democracy compel us to ensure that all citizens are numerate. If the conversation includes all disciplines rather than only mathematics, then QL can be seen not as a lower track within mathematics but as a way to extend mathematics into areas that many students may find more rewarding. Structures designed to sustain a QL conversation that is genuinely inclusive of all disciplines can not only help faculty address the numeracy needs of students, but also help to integrate mathematics into the fabric of college or university academic life.

Quantitative Literacy in K–12 Education

Mathematics courses that concentrate on teaching algorithms, but not on varied applications in context, are unlikely to develop quantitative literacy.
—Deborah Hughes Hallett

High schools have not taken on the task of developing courses specifically aimed at teaching a kind of practical, context-based "political arithmetic" that would help students learn to evaluate the kinds of numbers that are routinely invoked in political life.
—Patricia Cline Cohen

Paradoxically, students are being trained to perform operations that can now be dealt with using suitably chosen software, but all too often students share the computer's inability to analyze a problem and to reason.
—Geoffrey Howson

From one perspective, school mathematics is a fertile field for quantitative literacy. Students demand and teachers provide copious examples of how mathematics supposedly can be used. Students in both branches of a two-track system (see sidebar) are offered mathematics that could be of use, the major difference being the time to payoff: "academic" students are supposedly preparing for college programs that build on mathematics while students in "general" or "vocational" tracks are supposedly preparing for more immediate employment and life skills (e.g., personal finance). The NCTM Standards[50] reflect the increased need for all students to study mathematics that is not only intellectually important but also demonstrably useful. To enhance utility, topics from probability, statistics, and discrete mathematics have been added to mathematics standards to supplement traditional high school emphases on algebra and geometry. Ideally, according to this plan, any student who finishes a modestly strong high school mathematics program would graduate with all the fundamental QL skills needed for work or further study.

Many claim that this strategy is not working. The way children are taught to add fractions provides a clue: first find the lowest common denominator, then convert all fractions to that denominator, then add the numerators, and finally, reduce the answer to lowest terms. "Nobody does that outside of the schoolroom," observes Johns Hopkins economist Arnold Packer. "Imagine a school cafeteria where the selected items totaled three quarters and three dollars and four dimes. The schoolroom method would be to change all of these in for nickels. Or go to the shop. Maybe the problem is adding one foot and 8 and 1/16 inches to six feet and 1/4 inches. Would any carpenter change it all into sixteenths? It is a very rare situation when one needs this [school] method, say to add odd fractions, like fifths and sevenths."

In fact, says Project 2061's former director George Nelson, "school mathematics is typically formal and theoretical, thus not a welcoming environment for quantitative literacy." As every child knows, there is a big difference between authentic "real-world" problems and their classroom versions—boats crossing a river, trains meeting in the night, flavors mixing in punch bowls ... (Physicist Richard Feynman once ridiculed a school exercise that asked students to add up the brightnesses of a half-dozen stars—as if that were a meaningful application.)[51] Examples like these help explain the many studies that show that despite standards rich in the language of quantitative literacy, the majority of U.S. students leave school quantitatively illiterate.

There is another factor that is perhaps even more important: teachers' innumeracy. "To a considerable extent, writes Yale mathematician Roger Howe, "today's curriculum is more or less what today's teachers can deliver." Recommendations to improve quantitative literacy that ignore the need for capacity building among teachers—especially elementary school teachers—

The Dilemma of Tracks

The strongest cultural force shaping school mathematics is the tradition of tracking, which is especially widespread in high school. The college-preparatory track has preparation for calculus as its goal and does not include significant contextual uses of the mathematics. Measurement, geometry, data analysis, and probability—all parts of most school mathematics curricula—have strong QL themes, but with calculus as the goal these get short-changed. Thus by attempting to articulate well with colleges, schools narrow the coverage of mathematics to what is needed to succeed in calculus. The majority of high school students—about three out of four—who never make it through a calculus course never reap the benefits of this narrowed mathematics curriculum.

On the other hand, students who are in a non-college preparatory mathematics track are often short-changed by the lower level of the courses, and find themselves unprepared for college mathematics. When they arrive in college, as many do, they are likely to enter the wasteland of remedial courses.

—Bernard Madison

are unlikely to be effective. “The capacity of the teaching corps is not a peripheral issue, to be resolved after formulation of the ideal curriculum,” observes Howe. “It is a central issue.”

Because of their education and training, many teachers are not prepared for or comfortable with the mathematics required for quantitative literacy. “Teachers tend to teach what they were taught, imitating the way they were taught,” writes high school mathematics teacher J. T. Sutcliffe. “Those who are less confident with mathematics tend to focus on skills, especially skills separated from a meaningful context that would support quantitative literacy.” And what is true of mathematics teachers is even more true of teachers from other disciplines. According to Johnny Lott, former president of NCTM, it is simply unrealistic to expect that teachers of other subjects will either know or understand what might be considered quantitative literacy. The costs of a teacher corps with low numeracy skills are huge. They include, according to Howe, “extensive remediation, low achievement, inadequate skilled labor, and impoverished political discourse.”

Context

Almost without exception, everyone who engages the issue of quantitative literacy concludes that “in context” is one of its defining features. The essence of QL is to use mathematical and logical thinking *in context.* This is why statistics is cited more often than calculus as a good course in which to learn numeracy. Statistician Scheaffer calls context “the very lifeblood of statistics.” Most scientists and many mathematicians also affirm the importance of context (see Richardson & McCallum, cited above). Johns Hopkins economist Arnold Packer carries this argument a step further, urging that context replace what he calls “x,y mathematics” throughout K–12 education.

Yet many scientists and educators distinguish between using mathematics in context and teaching mathematics in context. Even many applied mathematicians worry that teaching mathematics in context is often impractical, unnecessary, or misleading. What is context to one student will often appear as confusion to another; moreover, it is often easier to apply principles and skills after they have been learned than before. Too much teaching in context can lead to a loss of general principles, thus depriving students of one of mathematics’ major strengths (see sidebar).

The debate about teaching general principles or specific contexts divides faculty both within mathematics and across disciplines. Whereas QL requires context, mathematics is built on general principles. Yet paradoxically, to

Consequences of Contextualization

"Mathematics instruction should be contextualized and avoid the abstraction associated with the traditional curriculum." This common refrain of current reforms is more complex than most of its advocates appreciate. One argument, which goes back to Dewey and others, is that learning best starts with experience, to provide both meaning and motivation for the more general and structured ideas that will follow. Dewey's idea differs in two respects from the above recommendation. First, it does not eschew abstraction. Second, it speaks of the experience of the learner, not of the eventual context of application of the ideas, which may be highly specialized and much later in adult experience, still remote from the learner's.

Another argument is that mathematics is best learned in the complex contexts in which it is most significantly used. This idea has a certain appeal, provided that it is kept in balance. Authentic contexts are complex and idiosyncratic. Which contexts should one choose for a curriculum? Their very complexity often buries the mathematical ideas in other features, so that, while the mathematical effects might be appreciated, there is limited opportunity to learn the underlying mathematical principles.

So the main danger here is the impulse to convert a major part of the curriculum to this form of instruction. The resulting loss of learning of general (abstract) principles then may, if neglected, deprive the learner of the foundation necessary for recognizing how the same mathematics witnessed in one context in fact applies to many others.

—Hyman Bass

improve quantitative literacy we have to wrestle with the difficult task of getting students to analyze novel situations, which requires understanding general principles. This kind of teaching is much harder than teaching a new procedure; it strikes at the heart of what students believe mathematics to be. Harvard psychologist Howard Gardner confirms this impediment:

> Perhaps the greatest difficulty in the whole area of mathematics concerns students' misapprehension of what is actually at stake when they are posed a problem.… [S]tudents are nearly always searching for [how] to follow the algorithm… Seeing mathematics as a way of understanding the world is a rare occurrence.[52]

Unfortunately, there are strong pressures on mathematics instructors to use teaching practices that are diametrically opposed to those that promote the kind of understanding that Gardner seeks.

Because of the ever-present possibility of unanticipated side effects, it is important to examine and anticipate the impact that a viable QL program might have on students, especially those often termed "at-risk." On the one hand, QL may open up new routes to advanced learning and productive careers that are now blocked by the filtering power of the traditional algebra-calculus sequence. On the other hand, so long as prestige remains with the traditional courses, many at-risk students (and their parents) may imagine QL as a program for less able students and thus resist the risk of being stigmatized by that perception. The tradition of tracking in mathematics education—much stronger in the U.S. than in some other countries—introduces a special complication for strategies designed to enhance QL.

QL and Technology

With rare exceptions, adults' quantitative encounters are mediated by technology: calculators for routine arithmetic; spreadsheets for more complex tasks; software for preparing taxes and managing investments; computer generated graphs in news; etc. One would think that such extensive engagement with technology would naturally enhance quantitative capabilities. But many adults—perhaps most—believe otherwise. Contrary to the prophets who promised an educational bonanza from the digital age, pundits and educators often argue that technology is a major contributor to quantitative illiteracy.

Actually, public reactions to technology show a strange dichotomy—opposition to calculators in elementary school combined with support for computers in high school. But this split opinion may not be so strange after all; it may only reflect the fact that the only mathematics adults are sure is useful is the arithmetic that they learned in elementary school. Most adults feel they would have lost something important on which their independence depends if they didn't carry school arithmetic around in their heads—however rusty and dysfunctional it may be. The independence of thought afforded by school arithmetic is not unlike the autonomy associated with driving a car.

Very few adults have the same kind of feeling about algebra, however. We all use school arithmetic—but not algebra—all our lives, so arithmetic has a sense of validity and necessity that higher mathematics lacks. We also know from experience that whatever degree of quantitative literacy we may have has been built on early rather than later learning of mathematics. So adults' resistance to calculators in elementary school may come from a sense of potential loss for what they had gained without technological aid in those early grades, while the lack of resistance with regard to what happens thereafter derives from the lack of any sense of useful gain in later grades.

Transition from School to College

> A principal cause of the transition problems in U.S. mathematics education is the lack of an intellectually coherent vision of mathematics among professionals responsible for mathematics education.
>
> —Bernard Madison

> In 1995 universities in the southeastern United States devised 125 combinations of 75 different placement tests, with scant regard to secondary school standards. —Michael Kirst

> Paradoxically, the fastest growing part of the high school curriculum is college level study, while the biggest part of our collegiate mathematics program is remedial, that is, high school course work. It is not clear that either of us does the other's work particularly well.
>
> —Janice Somerville

Educational transitions are a major sources of stress for students, especially in the discipline of mathematics and the domain of quantitative literacy. College instructors in the mathematical sciences face classes filled with students whose quantitative preparations are enormously varied, many quite weak, and whose readiness to make use of earlier quantitative experiences is barely discernable. Faculty in other disciplines that depend on mathematical and quantitative tools face the same challenges but have much less time to devote to diagnosis or remedy. For them, all too often, avoiding quantitative issues is the easiest option.

Early in the last century, schools and colleges contended over issues of content and subject-matter standards as each sought in turn to exercise leadership in the transition from upper secondary school to the first college years. After WWII, the focus on coordination of academic standards was displaced by an emphasis on "aptitude" exemplified by what was then called the Scholastic Aptitude Test (SAT) (and which is now called only "SAT" without any reference to aptitude). Test scores gradually replaced subject-matter standards as dominant criteria for college admission. At the same time, secondary schools added nonacademic electives for those students who did not intend to prepare for college. Today, school and college faculty belong to separate professional organizations that reduce professional interaction. Policymakers also work in separate orbits that rarely interact.

Michael Kirst, education policy expert at Stanford University, describes today's high school curriculum as "unmoored from any continuous vision of quantitative literacy." School-college articulation, he argues, is focused on access rather than academic preparation, in part because access is the "theme" of the many professionals who mediate between secondary schools and universities (e.g., high school counselors, college recruiters, college admissions and financial aid officers). In recent years, state-directed efforts have enunciated standards with two interrelated goals: clarifying what students must know and be able to do in grades K–12 and then aligning these standards with assessments, textbook selection, and accountability measures. However, these reforms have ignored the lack of coherence in content and assessment standards between K–12 and higher education. Until educators address this issue, observes Kirst, secondary schools and their students will have no clear sense of what knowledge and skills constitute mathematical or quantitative literacy. And with no clear goals or directions, no school can develop "any authentic means by which to judge its progress."

Teachers and parents are also confused. They receive mixed signals about the relative importance of traditional mathematics and quantitative literacy. Janice Somerville of NASH takes mathematics departments to task for ignoring this issue (see sidebar). It is students, notes Somerville, who pay the price for incoherence in educational goals.

In the United States, most educational policy that matters is made at the state and local levels. The K–12 sector spans nearly 20,000 districts and 80,000 schools, each subject to layers of local, state, and (increasingly) federal mandates. Higher education, as Arkansas mathematician Madison describes it, operates as 3000 or so "independent contractors" who "unwittingly" wield considerable influence on K–12 education—on parents and students through coveted spots in freshmen classes and on curricula through the influence of the academic disciplines. Although some such as NCTM's Lott argue that QL education must begin long before high school "else it be doomed to failure," QL is largely a secondary and postsecondary concern. According to Michael Kirst, policy disjunctures between K–12 and higher education concerning QL will be hard to mend in the absence of a national center and institutions in each state whose mission is K–16 alignment and reform. Nonetheless, he warns about the dangers of centralization: "With more specificity comes less flexibility."

What do Mathematicians Want?

I am puzzled by the relative silence in these discussions about ... the place of QL in mathematics departments.... Consider a little data that provides a national perspective on the flow of students from secondary school to higher education.... Today, three out of every four high school graduates go on to postsecondary education within 2 years—45% in four-year colleges, 26% in two-year colleges, and 4% in other postsecondary institutions.

But as we all know, many first year students are not fully prepared for higher education. Nearly 30% must take remedial courses when they enter college, and most of those (24%) are in mathematics. Moreover, remediation is most heavily concentrated in colleges with high minority enrollments: in these institutions, 43% of entering students require some remediation, 35% in mathematics. Not surprisingly, students who require extensive remediation graduate at significantly lower rates than other students: In fact, those needing three or more remedial courses graduate at one third the rate of students who enter college fully prepared.

Clearly, students and parents have understood that in today's world virtually all students need to graduate from high school prepared to continue postsecondary education. Indeed, as these data show, students are entering college in ever larger numbers, but they are clearly not mathematically prepared at an appropriate level. I would argue that the human and financial costs of the disconnect between standards for high school graduation and what we require for college readiness is no longer tolerable.

— Janice Somerville

Because the transition from secondary to postsecondary education poses so many difficulties for students, many policy leaders now urge that educators plan from a K–16 perspective. Typically, courses are the currency in which articulation is measured, so K–16 planning agreements usually specify course content as a way of ensuring cross-institutional transparency. But since QL is rarely explicit in courses—it functions best "across the curriculum"—and since it depends on contexts that will likely differ from one campus to another, course-by-course articulation agreements often work against efforts to achieve quantitative literacy.

As a consequence, some states are beginning to explore higher-level agreements based on "competencies" rather than courses. It is too early to judge whether these approaches will work better than course specifications, but almost certainly they will require the development of goals and instruments for assessing student learning. In the absence of clear consensus on levels of QL appropriate for each educational level, educators are proceeding with caution. The wide variety of approaches used by colleges and universities to develop a QL standard (see Part II for some examples) signals on the one hand the start of a movement in higher education to take QL seriously as an objective of learning and, on the other hand, the diverse and often rudimentary expectations that colleges believe to be realistic for current students.

Many complex issues about articulation appear when one considers mathematics education from a K–16 perspective. It is one thing to think about QL horizontally across the curriculum, but quite another to conceive of it vertically as one is compelled to do when considering the mathematics curriculum. Difficulties caused by inadequate articulation can have grave consequences for students. Mathematics is a gatekeeper to many mainstream careers, so lack of transparency in the transition from secondary school to college will differentially handicap students who may not have other means for discerning the most appropriate educational paths. High school juniors and seniors are focused on tests required for high school graduation and college admission, not on college placement exams and undergraduate general education requirements. Many students do not realize the importance of taking mathematics in their senior year as part of their preparation for college. One consequence is high enrollments in remedial courses, much of it unnecessary. As Madison notes, "having to repeat work, not making progress toward a degree, and studying uninspiring subject matter makes remedial mathematics courses unusually dreary."

Several large projects are underway to explore these issues. One, *Quality in Undergraduate Education* (QUE) seeks to define expectations for college students in mathematics and several other disciplines at three points in a student's undergraduate program (entrance, transition from sophomore to junior, and graduation). Another, *Standards for Success*, seeks to define common expectations in mathematics for students entering higher education in two cohorts—one level for all students, and a higher level for those planning on a program of study that will require college-level mathematics. (These two projects are described briefly in sidebars.)

One indication of the current complexities of the school-college transition can be seen in a recent national survey that estimates that half (approximate-

Quality in Undergraduate Education (QUE)

A project of the Education Trust and the National Association of System Heads (NASH) in association with Georgia State University, QUE is a national project of faculty at selected four-year public institutions and their partner two-year colleges who are establishing draft, voluntary discipline-based standards or outcomes for student learning in the undergraduate major. Objectives of QUE include development and use of standards for transfer from 2-year to 4-year college and development and use of standards for graduation from college. Further details can be found at www.gsu.edu/que. (QUE is funded by grants from The Pew Charitable Trusts and the ExxonMobil Foundation.)

Standards for Success

A project of the American Association of Universities (AAU), Standards for Success is intended to influence high schools by establishing and disseminating a consensus statement on the key knowledge and skills college freshmen need to be successful at research universities as well as to create linkages between these indicators and state standard-setting activities. The project operates from two centers: the University of Oregon is coordinating preparation of the consensus statement on standards for success, while the center at Stanford University is conducting a detailed analysis of state standards and assessment systems. Further details can be found at www.s4s.org. (Standards for Success is sponsored in part by a grant from The Pew Charitable Trusts.)

ly 3.5 million) of all juniors and seniors in U.S. high schools are enrolled in courses that carry credit for *both* high school graduation and college degrees.[53] A new joint committee of three mathematics societies dealing with mathematics in secondary school, two-year colleges, and four-year colleges is beginning to study issues involved with articulation at this level.[54] This committee has identified eight areas of concern, each of which intimately affects options for achieving quantitative literacy:

Overlapping Programs. The majority of enrollments in collegiate mathematics (over 60% in four-year colleges and universities, approximately 80% in two-year colleges) are in courses whose content is taught in high schools,[55] while the fastest growing part of the high school mathematics curriculum is in courses that carry college credit. [56] Much of this growth has been fueled by marketing of Advanced Placement (AP) courses,[57] notwithstanding recent studies that question the efficacy of these programs.[58, 59]

Transition Testing. Mathematics testing at the school-college interface—high school exit, college entrance, college placement, and rising junior—is often inconsistent.[60]

Transfer of Credit. Transferability of credit among colleges is an issue of growing importance because of student mobility and the wide variety of course delivery systems.

Mathematical Preparation for College. Few states make clear to high school students the mathematical expectations of higher education, and those few exhibit considerable variety in form, content, and level of detail.[61]

Curricular and Pedagogical Alignment. The variety of content and instructional methods within both high school and college mathematics makes alignment complex and frustrates students.[62]

Competing Pressures. Mathematics departments, both in high school and college, are subject to a variety of pressures that are often inconsistent, including state or district standards, mandated tests, and demands to prepare students for work and for college.

Articulation Across Disciplines. Mathematics and statistics are taught and used in many courses in school and college, yet there is rarely any coordination with its use in various disciplinary courses.

Articulation Between Tracks. Many mathematics programs offer different tracks in mathematics aimed at different student destinations. However, transfer from one of these tracks to another is usually difficult.

College admission practices play a key role in suppressing innovation in quantitative literacy at the K–12 level. "The need to develop quantitative literacy will only be taken seriously when it is a prerequisite for college," writes Arizona's Deborah Hughes Hallett. "Students and parents are rightfully skeptical that colleges think probability and statistics are important when they are not required for entrance. Making quantitative reasoning a factor in college admissions would give QL a significant boost."

A focus on expectations for college entrance would catapult QL into the national debate about other indicators of high standards such as SAT and Advanced Placement (AP). A corollary would be that school administrators and college deans would begin asking faculty what they are doing to advance quantitative literacy for all students and in all majors. Mathematics departments, especially, need to ask whether students who enroll in their courses are achieving the quantitative skills necessary to enhance their and the nation's quality of life.

Assessing Quantitative Literacy

> Fixing assignments and tests may be far more important than fixing syllabi.
> —Peter Ewell

> We do not really know if we are making progress [since] … we do not have genuine benchmarks for what constitutes quantitative literacy.
> —Rita Colwell

> Assessment of QL requires tasks that challenge the learner's judgment, not just exercises that cue the learner.
> —Grant Wiggins

From one perspective, the accelerating movement for educational assessment and accountability has enhanced interest in QL by exposing to public scrutiny the flaws of our entrenched system of mathematics education. Nonetheless, QL proponents generally do not view assessment experts as allies because they fear that test developers will distort QL to fit preconceived ("off-the-shelf") assessments. Since the nation's testing regime is so centered on mathematics (and reading), broad movement towards QL would require significant change in the nature of questions that are asked on these important exams. Colleges and universities rely on the SAT and ACT in order to obtain some uniform national measure, but neither of these tests is designed to assess quantitative literacy.

Because test items posed under standardized conditions are generally decontextualized by design, meaningful assessment of QL challenges the very conception of "test" as that term is ordinarily understood. Context matters for QL, so it must come to matter in QL assessment. Yet if context is by definition unique, asks testing expert Grant Wiggins, can we ever have standardized tests "in context"? "The consequences of one's work need to be felt in context in the same way they are on the playing field or in the auditorium." For Wiggins, the most effective QL assessments would be those in which students directly experience the actual effects of their work—as happens, for example, in writing computer programs. Developing such assessments for QL is indeed a major challenge.

Further, Wiggins argues, QL forces students to confront situations that have no signs suggesting which path may lead to a solution. This raises havoc with traditional psychometrics because it is the exact opposite of an effective test item. Assessment, Wiggins argues, should be "built upon a foundation of realistic tasks, not proxies." It should be "faithful to how mathematics is actually practiced when real people are challenged by problems involving numeracy."

Decontextualized items are one thing; deconceptualized items are another. Both populate standardized tests. "Most state exams, end-of-course exams, and even the ACT and SAT mathematics exams have very few questions that focus on the ability of students to reason and think with mathematics to solve authentic real-world problems," observes Eugene Bottoms, Director of the High Schools at Work program of the Southern Regional Education Board (SREB). "In too many instances, mathematics exams encourage teachers to teach the wrong way. They encourage teachers to cover the material, teach students the procedures, and hope that they will remember them long enough to pass the exam. The emphasis of the exams, and thus of teaching, is not on deep understanding of mathematical concepts or on advancing students' reasoning skills." To gauge quantitative literacy, assessments must seek evidence not just of decoding ability but also of meaning-making.

There is yet another reason why it is important for assessments to emphasize context, concepts, and meaning: tests determine what teachers teach and what students learn. "Many teachers teach only what will be assessed on high stakes exams," writes J. T. Sutcliffe. "Since problem solving and reasoning are seldom assessed on such exams, it stands to reason that problem solving and reasoning are seldom taught."

Michael Kirst notes in his review of state policies that recent standards-based reforms have ignored the lack of coherence in content and assessment standards between K–12 and higher education. "Until educators address this issue," he argues, "secondary schools and their students will have no clear sense of what knowledge and skills constitute mathematics literacy." One particular: the content of mathematics placement tests in college continues to reflect (however imperfectly) their original use, namely, to determine where to place students on the rungs of the ladder leading to calculus. But, as Janice Somerville points out, many public systems of higher education now use performance on these placement exams to determine whether a student can begin credit-bearing courses or even whether the student can enter a four-year college at all. Not only is QL lacking from the objectives of this system, but so is any rational relationship between means and ends.

On campuses, the critical issue is how to determine the degree of QL competence of students in a manner that is both constructive and reliable. "We must not lose sight of the fact that our goal is student learning," write Arizona geophysicist Randall Richardson and mathematician William McCallum. "It is far too easy in the heat of battle over establishing quantitative literacy requirements, setting up support centers, or revising our individual courses, to forget that the student must be the focus of our efforts. The question of 'what works best' must be answered in terms of student learning. In order to do this, we must establish clearly defined student learning outcomes in quantitative literacy. We must be able to develop measures for these outcomes as part of an ongoing assessment program. Key to the success of such an assessment program is feedback into the way we are teaching quantitative literacy. Without such formative assessment, debates on how to improve quantitative literacy will be driven by anecdotal experience and the force of individual personality. Students deserve better."

Clearly quantitative literacy must develop accepted methods of assessment if it is to achieve its goals. Assessment of QL is important to inform public policy, since only evidence of student progress will convince policy-makers that both the objective and the means are worth the cost. What parents, business leaders, and policy makers really want is not only better performance on mathematics tests but also increased capacity to meet the quantitative demands of work and life. They call for more advanced mathematics not because they really value the formulas and procedures of trigonometry or calculus but because that is the closest proxy they see for QL. Unfortunately, despite some slow improvement, current standardized tests are unlikely to help much in advancing the goal of QL.

The chief policy goal for assessing QL should be to move in the direction suggested by NSF Director Colwell: establish proficiency levels for QL that relate to students progress through K–16 education. Ideally, these proficiency levels would be tied to existing and emerging state standards in mathematics and science, to college expectations for entrance and placement, and to criteria for "rising junior" status and graduation. The goal would be a common QL yardstick benchmarked to the several stages of education, including preparation of prospective teachers in all fields.

Challenges for Policy and Practice

> There is no agreement, nor even a forum to deliberate about a possible consensus, concerning mathematical or quantitative literacy within the United States' K-16 education system. —Michael Kirst

> Our knowledge base about QL is not yet adequate for designing major interventions in the school curriculum. —Hyman Bass

> QL will not thrive when viewed as a part of mathematics or statistics. It must be viewed as a pursuit in its own right. — Phil Mahler

> Lack of support from administration and minimized opportunities to network with fellow faculty members often leads to demise of new curricula. —Susan Ganter

Not surprisingly, many challenges must be overcome for any cross-disciplinary enterprise to gain a foothold in a college curriculum that is encumbered by tradition and dominated by disciplines. For quantitative literacy, these include:

- The dominance of College Algebra as the customary way of fulfilling a QL requirement in the "check off" curriculum.
- The reluctance of mathematicians to play a leadership role in developing a QL-oriented curriculum.
- The habit in many disciplines to avoid quantitative content because of student innumeracy.
- Lack of experience in understanding and assessing student progress in QL.
- Inappropriate use of placement tests as surrogate assessments of QL.
- School-to-college articulation policies that are driven primarily by preparation for calculus.
- The challenge of thinking about QL and mathematics effectively in a K-16 context.
- Translating QL into a policy lexicon and scenario that is clear and compelling.

- Issues of scale to help QL escape the destiny awaiting most small, marginal innovations.
- The dilemma of tracks, of the nearly irresistible tendency to see QL as a lower track for weaker students.
- The widespread assumption, especially among mathematicians, that students in the traditional calculus track automatically become quantitatively literate.

Clearly there is much yet to be learned about quantitative literacy. Some are issues of fact to be answered by research; others are issues of intention to be arrived at through a process of deliberation and consensus. These include issues of status, goals, practices, assessment, and pedagogy:

Current status of QL:

- What is the QL profile of current high school graduates? Of two-year college graduates? Of four-year college graduates?
- In what aspects of QL are students well or ill prepared when they transition from school to college and from college to employment?
- What do employers need in terms of the QL preparation of new hires? What have they found in recent years?
- Which states include QL as part of their diploma or degree requirements? What are their definitions and benchmarks?
- Do any disciplines have guidelines for QL?
- How, if at all, do colleges express their QL goals?
- What strategies do colleges employ to encourage (or require) students to meet QL goals?
- To what degree and in what depth do states, districts, and colleges assess QL goals?

Goals and benchmarks:

- What are appropriate goals for QL at different educational levels—secondary school, two-year college, four-year college, master's level? What benchmarks demarcate these different levels?
- In what ways do QL expectations become more sophisticated as students move through school and college? How does QL in college differ from QL in high school?
- In what ways do experiences with QL enhance students' subsequent learning of the mathematical, natural, social, and behavioral sciences? In what ways is innumeracy a handicap?

Collegiate policies and practices:

- What criteria do colleges use to determine which courses satisfy a QL requirement? Do such special courses accomplish their goal?
- What strategies do colleges use, if any, to ensure that students' numeracy rises continuously throughout their undergraduate studies?
- Can QL be effectively embedded across the disciplines (as writing generally has been)?
- How can colleges ensure that QL does not become a lower track?
- Where does QL fit into teacher education (both at the elementary and secondary levels)?
- Is the reported innumeracy among college graduates primarily because colleges have not made an effort in this area, or because the efforts they have made are ineffective?

Assessment and testing:

- Can we identify assessable levels of QL for key educational transitions (e.g., grades 12, 14, 16)?
- How well do widely used standardized tests assess students' quantitative literacy?
- How can real world circumstances best be simulated for QL assessment?
- Do colleges assess the QL of their students? If so, how is it done? What are their findings?
- Do we know if students leave college more numerate than when they enter?
- To what extent does quantitative literacy correlate with other established measures of student success?

Pedagogy

- Which pedagogical practices promote quantitative literacy?
- How do high-stakes standardized state tests influence the role of QL in K–12 instruction?
- Which K–12 and higher education policies promote QL-friendly pedagogy?

Clearly these questions are interconnected with each other and directly impact the fundamental issues we have been discussing: why QL is important, how QL can be assessed, and how students' QL capacities and performance can be improved. The final section of this report contains suggestions for addressing these important issues.

Achieving Quantitative Literacy

> If we want students to know something, we should teach it.
> — Hyman Bass

> Most Americans are unaware of how mathematics permeates their lives. We must find ways of bringing their daily quantitative activities into focus.
> — Rita Collwell

> The remedy for the widening gulf between those who are literate in mathematics and science and those who are not is democratization—making mathematics and science more accessible and responsive...to the needs of all citizens.
> — Anthony Carnevale

The need for education that promotes quantitative literacy is new, critical, and growing. One force creating demand for QL is the increasing role of computer-mediated data in all aspects of our lives. As the Internet makes accessible to average citizens information that had formerly been reserved for experts (e.g., doctors, bankers, scientists), a social movement has taken hold that shifts decision-making responsibility from experts to novices. This devolution of decision-making helps accelerate a slowly growing shift of educational priorities (especially in the mathematical sciences) from a focus on the few to "education for all." QL is part of this broad social movement, its goal being forthrightly "for all," not just for the few. Education for QL is intended to help all people direct their own lives and contribute to democratic societies.

Findings synthesized from the national forum on QL (and reported in depth in *Quantitative Literacy : Why Numeracy Matters for School and College*) are enumerated below, followed by several Recommended Responses.

Finding 1. **Preparation:** Most students finish their education ill prepared for the quantitative demands of contemporary life.

Finding 2. **Awareness:** The increasing importance of quantitative literacy is not sufficiently recognized by the public or by educational, political, and policy leaders.

Finding 3. Benchmarks: The lack of agreement on QL expectations at different levels of education makes it difficult to establish effective programs for QL education.

Finding 4. Assessment: QL is largely absent from our current systems of assessment and accountability.

Finding 5. Professional Support: Faculty in all disciplines need significant professional support in order for them to enhance the role of quantitative literacy in their courses.

Findings and Recommended Responses

Finding 1. Preparation: Most students finish their education ill prepared for the quantitative demands of contemporary life.

Responses:

- *Embed quantitative literacy in courses across the curriculum.* Because changes in the world have increased the importance of quantitative literacy, many subjects—and not just mathematics—need to take QL more seriously than in the past. Instructors must end the practice of evading relevant quantitative issues as a favor to students who find them difficult. This evasion improperly contributes both to students' disregard for mathematics—since they never see it used—and to public ignorance of the significance of QL to different fields. As faculty now see improved writing as a goal in most courses, so they should see quantitative literacy as one of the explicit educational goals in courses across the curriculum.
- *Ensure that each student's numeracy is reinforced and developed continuously throughout high school and college.* Since QL involves the application of quantitative reasoning skills in a variety of contexts, education must include quantitative thinking as an integral part of many different contexts. QL is not a requirement that can be completed in one course and then checked off; it is a quality of mind that must be continuously exercised and strengthened. To ensure development of students' capacity for engaging in QL analyses, curricula need to be planned with this goal in mind both across the curriculum and vertically from secondary school to college. Corresponding assessments need to monitor this growth.
- *Teach students in mathematics courses to think mathematically in context.* For most students, skills learned free of context are skills devoid of meaning and utility. The separation of number from measurement and of symbols from meaning has had a particularly negative impact on students'

quantitative literacy. Most numerate adults learned much of the QL they know outside the mathematics classroom. Although numeracy clings to specifics and is inseparable from context, mathematics by its nature requires students to rise above context. Nonetheless, mathematics education could make a stronger contribution to quantitative literacy than it now does if it were taught more consistently in ways that emphasize context.

- *Create alternative routes to advanced mathematics.* The traditional six-year mathematics program in grades 9–14 is not the only possible way to gain advanced experience in the mathematical sciences, but its formidable presence in the curriculum causes many able students to drop out of mathematics well before they have achieved either an effective level of numeracy or a useful command of mathematics. Alternative entrées to the mathematical sciences via statistics, computing, discrete mathematics, and other mathematical topics could help many more students achieve quantitative and mathematical literacy.

Finding 2. Awareness: The increasing importance of quantitative literacy is not sufficiently recognized by the public or by educational, political, and policy leaders.

Responses:

- *Publicize examples of quantitative literacy in a full range of contexts.* QL is a relatively new educational goal without a universally recognized definition. Thus it often gets lost in the shadow of mathematics or statistics, both of which have an established curricular presence. To maintain focus on QL it is important that the people who support education—parents, taxpayers, voters—begin to understand what quantitative literacy is, why it is important, and how it cuts across traditional disciplines.
- *Align quantitative literacy efforts with other educational initiatives.* As a literacy, QL cannot stand apart. It is integral to core curricula, interdisciplinary programs, teacher preparation, inquiry-based pedagogy, K–16 strategies, and many other educational programs. Indeed, as QL should be embedded in many different disciplines, so also should it be embedded in these and other educational programs.
- *Engage faculty from different disciplines in dialogue about the role of quantitative literacy in the goals and requirements of their institutions, in their disciplines, and in their courses.* Being quantitatively literate is a responsibility of citizenship and a standard of performance in many careers. From journalists and politicians to nurses and teachers, everyone

bears a responsibility for understanding and dealing intelligently with quantitative data they face in their day-to-day lives. Educational institutions, especially colleges and universities, need to incorporate this new responsibility in their goals and curricula.

- *Make the role of inference and quantitative evidence in issues of public importance a priority for education.* Traditionally, quantitative tools have been viewed as important for students in engineering, the natural sciences, and more recently in the behavioral and social sciences. But with the advent of computers, data have come to play a crucial role in major public debates that shape our society. It is important, therefore, that all students develop the core quantitative competencies required for comprehending public issues.
- *Seek the attention of media, politicians, and funding agencies.* To gain momentum, QL needs both rhetorical and financial support: it needs to become "the next thing." Colleges and universities should issue press reports to help move QL into the domain of public debate. Then QL will become part of political dialogue as educational standards and testing now are. QL adds nuance and skepticism to simplistic public debates; it can cause people to inquire into the nature of educational standards and assessments. Funders will then listen in and support promising initiatives.
- *Create an annual QL decathlon to interest students and to illustrate what QL means in practical terms.* Many students thrive on competition, not just in sports but also in academic contests. Since QL touches so many different fields, the metaphor of a decathlon seems apt. An annual competition with tasks selected from a variety of domains would be an effective way both to call attention to QL and to help set standards for exemplary performance. Several associations with different stakes in QL (e.g., MAA, AAC&U, APA, ASA) could jointly sponsor such a competition; the cooperation required to develop the event would itself lead to important QL-related spin-off projects among the sponsors.

Finding 3. Benchmarks: The lack of agreement on QL expectations at different levels of education makes it difficult to establish effective programs for QL education.

Responses:

- *Establish benchmarks for quantitative literacy appropriate for secondary school, two-year college, and four-year college graduates.* One of the challenges of encouraging a culture that supports quantitative literacy is the need for clarity about expectations at different educational stages. Although some argue that QL should ideally be completed by the end of secondary school, most recognize that as students move through post-secondary education their experiences with and capabilities for quantitative literacy mature. Moreover, quantitative literacy—like any literacy—will quickly atrophy if not used. Benchmarks would help educators clarify QL expectations for students in their courses and also help students by making clear what they need to aim for.

- *Expect quantitative literacy for graduation both from high school and college.* If colleges ask their majors and degree programs to set explicit goals for quantitative literacy, both for admission and for graduation, these goals would rather quickly influence secondary school expectations and, more slowly, state exams for high school graduation. Using these goals, college programs across the curriculum can develop explicit plans for enhancing students' quantitative skill beyond basic levels so that seniors are at a more advanced level than first year students. Moreover, external accreditation and program reviews should look to these indicators of QL for accountability.

Finding 4. Assessment: QL is largely absent from our current systems of assessment and accountability.

Responses:

- *Increase the "footprint" of QL on widely used tests, especially those employed by colleges for admissions and placement.* Testing influences what teachers teach, what students learn, and what students come to expect as goals for their education. College admissions and placement tests especially influence the focus of high school mathematics, while major state and national transition examinations in grades 10-13 exercise considerable leverage over priorities for students in the transition from secondary to postsecondary education. For QL to gain a foothold in education, it must first increase its presence on major examinations.
- *Minimize reliance on decontextualized standardized tests.* In recent years, many state and national assessments have been increasing the proportion of contextual items that can serve as a weak surrogate for quantitative literacy. Test developers, policy leaders, and state assessment directors can strengthen QL by enhancing the move towards authentic contextualized test items.
- *Provide resources to help teachers develop instruments that effectively assess quantitative literacy.* The major need for effective assessment is at

the classroom level, both in schools and colleges. And it is here that resources are especially thin. School teachers whose own mathematical education focused on template exercises with predictable solutions will need assistance in learning to develop and grade assessments of QL; college faculty whose interests and preparation are focused inwardly on their own academic discipline will need encouragement to engage a cross-cutting capability like QL.

- *Create an on-line resource of QL assessment tasks for faculty to draw on.* Because the importance of context to QL makes broad summative assessment problematic, the most effective QL assessments will be local—at the level of the classroom, program, or institution (college or school). But it is not easy to meaningfully assess a context-intensive literacy such as QL within reasonable time constraints in typical classroom situations. Examples of approaches to classroom-based assessment would make it easier for faculty to develop instruments that monitor students' progress toward QL objectives, thus making it more likely that they will do so.

Finding 5. Professional Support: Faculty in all disciplines need significant professional support in order for them to enhance the role of quantitative literacy in their courses.

Responses:

- *Create a structural home for quantitative literacy.* Since QL cuts across all disciplines, it is never anyone's top priority. Instead, it is everyone's orphan. If innovation in teaching QL is to keep up with the rapid increase in demand, QL will need an infrastructure that plays the same role as professional societies, departments, and research institutes. Campaigning for QL across the disciplines is important, but not likely to be effective unless there is a structural intervention that will press the case for QL in a sustained, single-minded manner within an academic environment that is dominated by established disciplines.
- *Develop a technical resource infrastructure to support the teaching and learning of quantitative literacy.* All teachers who are attempting to infuse numeracy into their courses, even teachers of mathematics or statistics, need a place to which they can turn for examples, strategies, data, and course materials. Moreover, they need the personal assistance and encouragement of other like-minded teachers. The Internet offers a valuable tool for such an effort, but to be fully effective electronic tools need to be supplemented by face-to-face meetings and sustained by individuals whose

job it is to maintain and expand the necessary infrastructure (for example, professional development centers).

- *Develop a web-based inventory of QL resources.* Dissemination of innovative and exemplary approaches to QL is especially important in an area lacking traditional teaching aids such as textbooks and professional societies. The Internet is the obvious medium for such dissemination and already hosts examples of similar efforts that support innovation in teaching calculus, statistics, probability, and other subjects.[63] Such an inventory would require both financial and practical support; it might include class-tested tasks and assessment items as well as case studies of innovative QL programs, requirements, courses, and "best practices."
- *Support professional development opportunities and networks.* Summer and weekend workshops are the primary vehicle for professional development for both school teachers and college faculty: by working together on common tasks, teachers build the capacity to introduce new ideas in the classroom. Lacking the professional structure provided by traditional disciplinary societies, QL especially needs the support of workshops and teacher networks.
- *Infuse quantitative literacy into teacher preparation.* The widespread consensus that QL must be taught across the curriculum (rather than only in mathematics or statistics classes) implies that all teachers should be prepared to teach the QL strategies appropriate to their discipline. For this to happen, preparation in QL and opportunities to relate QL to diverse subjects must become part of the preparation of teachers in all fields, especially in elementary education. Accrediting agencies can help by insisting that QL be one of the issues that teacher preparation programs address in their periodic reviews.

Part II

Inside, Outside: Dialogues on Mathematics, Technology, and Statistics

The following dialogues are edited composites of several actual discussions that took place by e-mail between mathematical insiders and skeptical outsiders following the December 2001 national forum "Quantitative Literacy: Why Numeracy Matters for Schools and Colleges." These dialogues are not so much about quantitative literacy itself as about important background issues that frame the context for QL in education.

On High School Mathematics

Outsider: Let me begin by saying that it is very unlikely that anyone would worry about articulation in its relationship to any subject other than mathematics. This is because only mathematicians think of their subject, educationally speaking, as rigidly "vertical." Even foreign language teachers acknowledge that immersion (being thrown in at the deep end) is a much better way to learn a language than following the school sequence. Gradually throughout the last century, every subject except mathematics renounced "verticality" both as a conception of their subject and as a basis on which to build a curriculum.

It seems to me and others who worry about curricula that it is very important to know whether mathematics is truly a vertical discipline. If it is, then there are severe constraints on curricular maneuverability where mathematics is concerned. Whereas one would say about other subjects that there are many ways to enter and many paths to take, curriculum planners would be compelled to say about mathematics that there is only one door to enter and a single, narrow way to proceed. And so it may be. But one may be skeptical, because the GATC sequence—geometry, algebra, trigonometry, calculus—was designed for a Newtonian universe and it now seems pretty clear that Newton didn't see everything there is to see in the cosmos.

Another implication of mathematics' verticality, if this is genuinely essential to its nature, is that it must be contrasted with quantitative literacy in an even more pronounced manner. For surely it is in the nature of QL to suffuse and spread across domains, and therefore there could be no possibility of QL ascending hand in hand with mathematics on its vertical climb.

My point is that if the GATC sequence is not largely a historical artifact or, as others have argued, a social tool for sorting students, and instead is indeed the one and only true way in mathematics, then we are all very hemmed in when addressing issues of articulation. But if there is good reason to question the verticality of mathematics, or even to substantially qualify what we mean by it, then the options for action are increased.

Insider: Mathematics research is far from vertical, having many roots and branches — in nature, in abstractions, and in applications. Twentieth century mathematics research blossomed in dozens of subfields with work in one mysterious to workers in other subfields. After their independent development, almost all of these subfields have been shown to be related. In commenting on this phenomena, former AMS president Arthur Jaffe said "This seamless web is not only breathtaking, it makes it impossible to be encyclopedic in describing recent research and application, confounding any overly simple organizational scheme.[64]

The multiply-rooted, independently-developed, and now converging structure of mathematics research differs greatly from the vertical structure of the GATC sequence in the school-college transition years. This verticality and resulting lack of student success creates a major educational logjam in large part because of educational policies rooted in deeply held beliefs of many in the mathematics fraternity.

Much of the GATC sequence aims at preparation for calculus and consists of learning skills that can be performed by technology. Mathematicians can't agree on what skills are essential to have or on how or whether technology can help; some even ban technology in lower division undergraduate courses. We know that our own algebraic skills served us well and we see students falter because of their poor algebraic skills. These experiences reinforce beliefs that underwrite the GATC stranglehold.

Outsider: You appear to say that although the GATC sequence is not right for many students, algebra is essential for anyone who is ever to do much of anything mathematical. I have heard this from others, that algebra is the "language of mathematics." I don't know if "language" is used here as an analogy, metaphor, or what; but if it is true, the really critical matter for articulation then would seem to be the infusion of algebra throughout the mathematics curriculum (whatever the curriculum happened to be) and not the retention of the GATC sequence. To my mind, this then changes the articulation issue from one of verticality to one of algebraic fluency and how it is best acquired. But perhaps one is not meant to take this phrase "the language of mathematics" quite so seriously.

Insider: Mathematics courses usually require the students to *do* mathematics. We don't have many courses *about* mathematics. Consequently, one has to have command of some logical system of objects and rules to do mathematics. However, a command of arithmetic, some geometry and measurement, and rudimentary algebra opens up several possible routes.

Fields medalist William Thurston once wrote that mathematics was both deep (vertical) and wide (horizontal), and that in its breadth it spread like a banyan tree, laying down new roots from each major branch.[65] Less colorfully, I would assert that it is in the essence of mathematics, and perhaps of no other discipline, that one can move logically from any beginning point to any other. This feature of mathematics, which derives from its logical structure, totally voids any claim that GATC is the *only* way to approach mathematics.

For example, if one wanted to focus school mathematics on issues of risk and data analysis, it would be entirely possible to make data, probability, and elementary statistics the "subject" of school mathematics for 2–3 years. Students would move from arithmetic to ratios to probabilities to graphing data to calculating slopes to finding areas by throwing darts,…, learning all the essentials of algebra, geometry, and trigonometry and some calculus as tools that emerge naturally in the process of learning about data.

Or if you were in a great books curriculum, you could begin in Euclidean geometry, cover much of algebra as a notation for geometrical relationships (squares, cubes,…), do some data analysis via geometric probability, touch on simple calculus as Newton did it, and end with linear transformations associated with computer visualization. Along the way, the tools of algebra would be teased out of the material and organized into useful knowledge by use of contemporary notation. One could do the same via computer science, focusing on algorithms and recursion, spreadsheets and simulations.

My point is that there are dozens of logically possible alternatives to GATC. They would differ, of course, in their pedagogical and motivational appeal. Each would teach what is important of GATC intrinsically in a "learn-by-doing" mode. However, all but GATC itself would clash with current tradition, posing insuperable difficulties for students who move, as large numbers do; for textbook selection, since there are few if any complete curricular alternatives available; for teachers, who are too accustomed to GATC; and for colleges, who would not know how to judge what students had learned. Thus the argument for GATC is almost entirely social, not mathematical.

Outsider: So the GATC sequence was not handed down from God to be forever preserved because of its divine nature. Someone should say this, loudly. This means that, from a conceptual and intellectual point of view, the sequence schools elect is open to discussion and choice. Volition comes into it, some degree of freedom, and so on. It becomes possible, even necessary, to think about change and reform, about doing things differently where conditions suggest they should be done differently.

The constraints on this are "social" in nature, not mathematical, although you and I point to different social factors. You rightly say that the mobility of the population almost dictates that we have a universal sequence (though this doesn't require it to be the GATC sequence). I point to mathematics as the instrument employed to maintain a meritocratic social structure (through the dubious equating of mathematics and "merit"). But either way this seems to me to make the larger point that the mathematics curriculum we have is not in "the nature of things" but, on the contrary, is a social construction; and, what we know about the nature of the "social" is that it is subject to change.

The need becomes then, as Dewey once said, to acquire as comprehensive and reasonable grasp as possible of the entire social situation so that action can be fitted to these circumstances. Some of these "circumstances" will argue for change and some will argue for keeping things the way they are. Our responsibility as moral beings, Dewey said, is to strike the right balance between the two in such a way that action is not paralyzed and we are not stuck with outdated ideas and tools.

On College Mathematics

Insider: A key assumption underlies most discussions of curricula among mathematicians, namely, that students in grades 8–13 can be divided into two camps: "calculus-intending," and not. For example, critics of the so-called "reformed" high school curricula generally acknowledge that these curricula are much better than what came before for everyone *except* calculus-intending students. Critics of college algebra say that there is virtually no reason for anyone who is not calculus-intending to take college algebra. Advocates of QL often say that QL is precisely what is needed by students who are not going to take calculus.

The presumption of this dichotomy seems so profoundly embedded in the culture of mathematics that it is hard to imagine anything that could change it. But I wonder how well defined "calculus-intending" really is? How much churning is there from year to year in the students who say they are calculus intending? To what degree is the intention in the student or in the parent or the teacher? Might individual intentions be largely ephemeral even while overall averages are predictable?

Outsider: For a century or more, the high school mathematics curriculum has been divided into "college-intending" and "other." The "other" category included "general math," "commercial math," and so on. In the last twenty years, "college-intending" became "calculus-intending" and everything in the "other" category was denigrated to the point that "calculus-intending" is now regarded as the only legitimate sequence (even though only a relatively small number of students complete the sequence). As I recall, attrition rates in the "college-intending" sequence average nearly 50% each year in grades 10–12. But the real question is how entrenched this calculus-intending sequence is in the culture of mathematics? Since the rise to prominence of calculus as a high school subject is a relatively recent phenomenon, I would say that it can't be as deeply entrenched as appearances may suggest.

Insider: It may help to look at this from a slightly different perspective. With respect to mathematics, the schools used to be X% college-intending and Y% other (general, consumer, vocational). In the 1980s NCTM began saying that Y should be near zero; at the same time, society began arguing that X should be near 100. So they all got together on state standards based on these compatible premises. But then the pressure for calculus split X into P and U, P being the prepared students who took calculus in high school or as first year students in college (e.g., those who are "on track") and U being those who entered college unprepared for college-level mathematics. Now U is something like two-thirds or three-quarters with P being only one-fourth or so.

This suggests a need to examine the difference between intentions and results, between the forward-looking perspective of an adolescent and the backward-looking perspective of an educational expert. On the one hand is widespread rhetoric that all students should be "college intending" (which as you say, has become the same as "calculus-intending"). Parents and students see things from this perspective. On the other hand are mathematicians who, looking retrospectively at the results, assume that only certain students are in fact calculus intending and that these special students can be identified early enough to make a choice between, say, mathematics and QL, or between rigorous traditional courses and a new "fuzzy" reform curriculum.

This creates an excruciating conflict of goals: so long as "college for all" is our national goal, how can anyone advise high school students to do other than what university mathematicians say they want? And what they are saying is if you want to succeed in college, prepare for calculus the traditional way, but if you don't care about college, then it is OK to take the newer courses with a QL flavor. What kind of choice is that to put in front of a student or parent?

Outsider: Let me tell you how this looks to "outsiders" like myself. I have been in repeated discussions in which members of the mathematics community seem to say that most students intending to go to college are not best served by preparing for and taking calculus. This is not a question of the "inferiority" of these students, but rather that their likely life destinations are better served by mathematical (and QL) preparations that are other and broader than that provided by the calculus track. But when it comes to operationalizing this belief in any way, these same members of the mathematics community begin to qualify, hedge, and back off in such a way that their prior disavowal of "calculus for all" is so undermined as to be withdrawn.

Now, to some of us outsiders this looks like a lack of moral courage. It is like historians saying that "facts" in the E.D. Hirsch sense are unimportant and then wringing their hands when students place the date of the War of 1812 as "somewhere between 1825–1850." But of course, the stakes for students are much higher where mathematics is concerned. Anyway, this is why Jan Somerville says bluntly to mathematicians, "say what you believe and act on it. Stop saying calculus is not all-important if you are then going to abide practices that make it all-important."

But Jan and I are, after all, outsiders, and I am afraid that we must be missing something. It can't only be a failure of moral nerve, if it is that at all. There must be something profound, and perhaps intractable, about this issue that non-mathematicians fail to understand. But I must say, it is very puzzling.

Insider: Many mathematicians consider the use of mathematics by non-scientists as trivial and not of their concern. They view the educational pipeline as a path that leads to advanced mathematics—beyond calculus. Calculus is just the gateway to that path, and it got that way because it is also the gateway to science and engineering. The very earliest calculus reform talk was about making discrete mathematics the gateway—especially for computer science. Yet calculus remains the gateway for most. The problem is not really calculus; calculus is just a proxy for mathematicians' view of the goal of university mathematics education: to produce scientists, engineers, and—most importantly—mathematicians. In the view of many mathematicians, engineers and scientists are a just a useful byproduct, but they actually provide the power that installed calculus and keeps it there.

The early national efforts on behalf of mathematics[66] focused on the mathematics pipeline. That's the first place the one-year half-life scenario (then from grade 8 on) became widely known. The research leadership in U.S. mathematics is still focused on that pipeline and on getting money to support research. Support for broader education programs is most often seen as eventually leading to more mathematics PhDs or more money for the Division of Mathematical Sciences at NSF. Both entrenched calculus and calculus reform are just proxies for this larger professional enterprise.

Unfortunately for QL, mathematicians control much of the college and university mathematics curricula. To succeed in getting better attention to QL, advocates will either have to convince mathematicians that it is a good strategy for them or they will have to find a different power base for QL.

Outsider: There is an assumption running through all this to the effect that you insiders know how educational policy with regard to mathematics comes to be made and enforced. In short, this policy is seemingly made for most students everywhere by the mathematics faculties of research universities. Now, I do not doubt that insiders know more about this than us outsiders, but this seems to be only what is sometimes called "tacit knowledge"—something nowhere written down in books, or fixed in legislation, but rather that comes to be understood only through actual experience in a community of practitioners. The rest of us are all outsiders and have no way of knowing what is going on, or why.

Insider: You're correct that this knowledge is tacit. It is not written down anywhere and although there are a few cracks in mathematicians' power structure, it still generally holds sway. For example, the recent report on the mathematical education of teachers[67] recommended major changes in the preparation of middle and elementary school teachers, but almost no changes (save adding a capstone course) in the preparation of secondary teachers. One of the editors explains this by saying it was a bow to the power structure of mathematics: they would get worried if anyone told them to change the courses for mathematics majors. That major in mathematics has great influence on school mathematics and on mathematics in grades 13–14.

One's reputation within the mathematics community is almost always made by publishing mathematics research papers. Yet only a very small percent of PhD's in mathematics ever publish a single research paper beyond their thesis, and fewer still publish several. This select group of active mathematicians—a few hundred at best—wield enormous but indirect influence over undergraduate and graduate mathematics by setting the tone to which other mathematicians aspire. Most rarely teach lower level mathematics courses, certainly not college algebra.

Outsider: If this is so, it seems to me that those of us concerned about students' education have some obligation to bring this invisible hand into the light of day. If, that is, the whole "calculus-intending" business actually emanates from a small number of research mathematicians with specialized and exclusive interests who make educational policy bearing heavily on all students, then someone should put the facts on the table.

Insider: This is the core of the articulation dilemma—what Jan de Lange refers to as "environmental articulation." It seems to me that there are two choices: either mathematics changes or someone else takes control of general education.

Outsider: That's fine if someone is (or can be) in control. However, the way educational policy gets made vis-à-vis mathematics appears very different if one looks, as Michael Kirst does, up from below. From down there at the grass roots, the making of educational policy looks like a jumble of competing interests that inevitably produces confusion all round. So which is it: the one dominant force pressing down from above or the babel of voices arising from below? Of course, one could say that the one view applies to higher education and the other to K–12; but we all know that the two are both said to be in play throughout a long stretch of the curriculum.

As things stand, I see two very different opinions. Either someone is in charge or not. Quite a difference in outlooks, if I interpret them correctly.

Insider: Kirst's view—a maze of policies, agencies, and other entities—is a view in the large. Indeed within this maze, decisions are made in some states to require college algebra of all college students. Yet the mathematics power structure determines what college algebra is. The maze folks don't have the information needed to delve deeper and challenge college algebra. Critical decisions about what mathematics to teach are made locally and that local group is under the strong influence of national attitudes—editors of research journals and leaders of the American Mathematical Society and of the National Academy of Sciences. Mathematicians are elitists and we all listen to elitists.

Thus it is some *thing,* not some *one,* that is in charge: a power elite whose prestige influences everyone lower on the food chain. There are similar situations within all disciplines. Look how long the Western Civilization course lasted even after it was clearly wrongheaded. Freshman writing courses stayed mechanical for years after they should have changed. All that stability has to come from some power structure. Mathematics just needs change more urgently because it is more extensive in school and college. I am not sure it is more archaic than others.

Outsider: Indeed, all disciplines are ruled by power elites. But as you say, "mathematics is more extensive in school and college." That's the nub of the matter. As we all know, the SAT has a lot of influence on the school-college transition. It is curious that many mathematicians care a lot about SAT-M and other mathematics examinations, but influential English faculty didn't give a hoot about SAT-V. The fact is that the influence of the mathematics power structure reaches downward in the curriculum in a way that doesn't happen nearly as much in other disciplines. Historians have no say whatsoever about the K–12 social studies curriculum, and until very recently scientists' influence on school science has been almost equally negligible.

So, the mathematics power elite is in an exceptional position vis-à-vis K–16 education, and I do not think that there is a simple answer to why this should be so. Clearly, the answer is not very much to be found in the nature of disciplines in a generalized sense. Sorry to say it, but I think a bigger part of the answer is that mathematics serves a social policy function—that is, as Tony Carnevale emphasized, it sorts students into some who go one way and others who go another.

Insider: There may be another (or an additional) plausible explanation: the nearly linear progression of mathematical curricula that is essentially absent in other subjects (except for foreign languages). College historians, chemists, and English professors do not depend on the high school preparation of incoming students nearly as much as do mathematicians. So maybe the answer is actually "in the nature of the disciplines."

Outsider: A century ago, when the modern disciplines entered the curriculum, they all said that they were based on a "linear progression" structure; but the fact is that they did so largely because the classical curriculum had a prescribed linear progression and, in contesting it, the new disciplines had to argue that they had the same rigorous, linear, upward reaching nature and, thus, they found it necessary to structure study in the discipline accordingly. There is much more involved here than I can safely get into, including the fact that education was then (and well into the 20th century) based on the movement of a "class" as a collective body, ever upward, from one year to the next.

But this vertical edifice pretty much crumbled in most disciplines under the weight of a massive horizontal branching process. A vivid case study of this is seen in the *Double Helix* [68] where it turns out that discovery came not from a linear forced march but from an agile skipping around, learning what one can, and simply ignoring and flanking many basic things that one cannot understand.

The question is how can it be that mathematics is so rigidly "vertical" when most other disciplines, after an early fling with verticality, have given it up? You suggest that this is due to the nature of the discipline, which would make it then ontologically like no other discipline (and, in effect, "supernatural"). But mathematicians have had a devil of a time explaining how this is so to other people. I suggest, on the other hand, that there may be a more genuinely naturalistic answer, namely, that social policy wants and needs mathematics to be "vertical" as a means of maintaining hierarchical structure within mass society. If this social policy were to change, then, might not mathematics (educationally speaking) become more like other disciplines? I think it probably would.

Insider: From a mathematician's perspective, your line of reasoning borders on revolution, although it is not unlike what some international mathematics educators argue (e.g., D'Ambrisio and, to a lesser extent, de Lange and Niss). The extent to which mathematics is inherently vertical or horizontal has been discussed some within the mathematics community. Notwithstanding Thurston's description of mathematics as more like a banyan tree than an oak[69]—with deep roots emanating from every branch—99% of mathematicians and mathematics teachers think of it as like an oak: one trunk with branches at the top. The only serious debate within mathematics is where the branches begin.

You seem to be suggesting that mathematics is more like a bush than a tree, or perhaps more like the Internet—everywhere connected but with no beginning or center—and that its apparent structure is determined by social convention and convenience rather than by inherent logical necessity. I fear that trying to convince a research mathematician that the prerequisite for mathematics courses are the result of social policy would be about as effective as trying to convince politicians to base their votes on evidence and reason.

Outsider: Instead of a "bush" I have in mind the image that Darwin gave us—"a tangled bank." This becomes ever more true, I think, as one moves from abstract conceptions of disciplines to actual educational arrangements. In the latter, I agree with Dewey that social determinants far outweigh intellectual ones in producing order. To shift to history for a moment, history in the schools is what society wants it to be. Historians know this and back off. Maybe mathematicians do the same, though you seem to see it otherwise.

On Technology and Mathematics

Outsider: I am again trying to understand the relationship between the role of technology in writing and in mathematics. For starters, it is clear that technology has transformed the teaching and learning of writing in the last twenty years; but it has not yet come close to doing the same for mathematics. Could it? Should it? Are the situations in any way comparable? At first, one might think that they aren't. The computer doesn't do anything for the writer that he or she still doesn't have to do for him/herself. It makes it faster and easier to do, but it doesn't eliminate the doing. Mathematicians worry about technology because it eliminates the need for the individual to do certain things and so perhaps to understand them. It is not only a labor-saving device, but also a bypass for the human mind.

I tend to be skeptical of the mathematicians' concern about technology where learning is concerned. I am so, I guess, because the pedagogy that mathematics teachers employed with me from elementary school through college was to make my mind (mathematically speaking) as machine-like as possible. I can't remember a single mathematics teacher who ever suggested that there was anything to understand about any mathematical subject or activity. I always suspected that there might be, but the possibility never came up. Sometimes I thought that it didn't come up because mathematics was so laborious and time-consuming (like writing) to do by hand with pencil and paper. If there weren't so much manual labor, I thought, maybe we could have a discussion about why we were doing this and indeed what the "this" is that we are doing.

This last, of course, is what has happened with the pedagogy of writing because of technology. It has provided the means and time for opening the whole writing process to discussion and understanding—at least in principle. So, I'm inclined to believe that technology could make a huge difference in mathematics education. It seems to me that it would provide a lot more opportunity for talk about the "what" and "why" of mathematics learning—something, as I say, so lacking in my own mathematics education that I scarcely knew until many years later that it was missing.

Insider: Mathematicians' concern about over-reliance on calculators in schools is predicated on the assumption that mathematics in college will require manual skills. College mathematics can be taught entirely with technological aids—both calculators and computers—but very few places do that. So long as college entrance and course exams (and perhaps also science courses) require a lot of manual calculation, dual preparation and dual testing (with and without calculator) will continue. The argument for continued emphasis on manual calculations in school is largely empirical, not mathematical.

The role of technology in teaching writing is fundamentally different from its role in teaching mathematics. In writing, technology is used to make the task of revision easier; in mathematics, it is used to make it possible to solve harder problems. Without technology, students can be expected to do only the simplest mathematics problems—3×3 matrices, simple integer coefficients. With technology, teachers can assign and student can do more complex (and often more realistic) problems. I doubt that essay and term paper assignments have similarly increased in sophistication or complexity as a result of the availability of word processing.

Outsider: I think that you are wrong about technology and writing, or at least not entirely right. Technology actually allows writing tasks to be much more complex and sophisticated—"harder," as you say is the case in mathematics. Because of the interplay of word processing, e-mail, and the Internet, instructors can give students writing assignments that would have been unimaginable twenty years ago. The technology makes many things possible (not simply easier) that were not doable previously; and, in consequence, both students and the instructor work harder. To say technology makes "the task of revision easier" barely scratches the surface of the change involved where writing is concerned.

Insider: You're right: My argument about technology was only about the surface feature of eased revision, whereas when one takes a broader view including web research and internet exchanges, writing assignments can benefit from enhanced complexity and realism in much the same way as do mathematics assignments. So let me try again.

Why the difference in mathematicians' enthusiasm for technology? I'd claim there is very little difference when it comes to the features of mathematics assignments that correspond to revision or web research. However, mathematics takes about 14 years to introduce all the pre-technology skills that for writing are completed in about 8 years—that is, basics of arithmetic (spelling), algebra (sentences), geometry (paragraphs), and calculus (essays). Both writing teachers and mathematics teachers embrace technology more or less equally after students have mastered these "basics," using technology to help students improve and master their skills, but neither sees technology as having fundamental value in helping students learn these "basics." From this perspective, the importance of mastering skills without use of technology is comparable in both writing and mathematics—e.g., to formulate grammatical sentences or to write correct equations—but in mathematics the introduction of these skills occurs much later than for writing. This may help explain why high school and college teachers of mathematics display a different attitude towards technology than do high school and college teachers of writing.

A second element is that computers are more capable of doing mathematics than they are of writing. Except for those programs that generate random aphorisms, computers really cannot compose sentences or paragraphs. But they can create the mathematical analog of a paragraph by solving equations or carrying out lengthy calculations, both arithmetical and logical. Mathematics teachers are thus faced with a dilemma that does not (yet) confront writing teachers: whether to limit students' access to technology as a means of ensuring that they will learn to do certain parts of mathematics without computer assistance, or to remove from their curricular goals those things that computers can do better than people can. Writing teachers do not have to face this choice, since computers cannot yet respond with an intelligent paragraph when asked to write about a particular subject.

Outsider: Your second point is surely true, though I know that there are writing "theorists" who would leap at the chance of debating it. I suspect, however, that mathematicians will not be able to hold on to this "dilemma" for much longer (after another turn of the generational wheel). I recently heard a writer of children's songs say that the level of sophistication of music that she wrote for a 14-year-old twenty years ago she would now employ in writing for an 8-year-old. This has to do almost entirely with the immersion in the new technologies that children are now bathed in from the crib upward. Much as they might want to, mathematicians will not be able to exclude children from having computers/calculators until they (mathematicians) are ready for the children to have them. For one thing, parents will know the practical importance of getting these technologies in the hands of children as early as possible. Think of the technological experience of the generation now about to become parents. I don't think that any attempted prohibition by mathematicians will mean much to them.

I don't know what to say about your first point. My impression is that the notion of "the basics before technology" is fast disappearing from writing pedagogy. If a line is drawn, it has more to do with the availability of technology than learning theory. And the computer makers, needing ever larger markets, cannot allow availability to be an issue. This may be like Dewey in his lab school having children learn how to weave, make butter, and the like. He, too, thought that these processes should be understood by everyone so as to not lose touch with the fundamentals of human intelligence. But what he could do in a small-scale haven simply was out of the question in the larger environment of public schools, where the social-technical forces in the surrounding environment could not be kept out.

In sum, the issues you mention (for both mathematics and writing) are real enough, but they are unlikely to remain issues much longer in a society totally dependent on the technologies over which educators are trying to exercise some degree of prohibition.

On Statistics

Outsider: Statistics 101 is required in college about as often as Calculus 101, but by different departments. Yet more students seem to find statistics to be a worthless requirement; certainly they tend to put it off as long as possible, and report later that they don't remember learning much of anything important. Calculus of course has its own problems, but even at its worst students tend to view calculus as a legitimate albeit onerous requirement. Indeed, virtually all departments that require calculus do so because it is a prerequisite for several core courses in the major sequence. Statistics, on the other hand, often sits on the sideline, being required for the major but not for courses in the major. This pattern lends credibility to students' beliefs that statistics is not really of central importance to the disciplines that require it.

If this analysis has any legitimacy, can one imagine making an argument that introductory statistics should be a prerequisite for the great majority of mainstream social science courses? Wouldn't this give statistics the same kind of political legitimacy (both on and off the campus) that calculus now has? It would certainly provide a major step to improving students' quantitative literacy.

Insider: Students might not remember "learning much of anything important" from the typical introduction to statistics, and they might not ever be required to use specific techniques, such as constructing a confidence interval, in their later courses. It would be a mistake, however, to conclude from this that the course lacked value for them. Statistics is quite different from other courses, particularly mathematics courses, that most students take. Probability is slippery, statistical inference can be a subtle concept, experimental design is at once both simple and profound. A student who takes a good introductory statistics course is exposed to (and one hopes, learns!) a different way of thinking. Awareness of and comfort with probability-based inference, accounting for confounded effects, and the like give the statistically educated student a substantial advantage over others in understanding the world, especially causal relationships.

I wish more faculty had half the understanding of association, correlation, and causation that I expect a student to show on a final exam in my statistics course! Faculty—not to mention others with less education—routinely draw inferences that are seen by the statistically literate to be transparently false. Maybe we should say to students "You should take Statistics 101 so that you don't continually embarrass yourself later in life as you discuss with others the social, political, and physical world around you."

Outsider: Statistics 101, like Calculus 101, now appears in at least two flavors—one focused on concepts, the other on traditional techniques. Many mathematicians and scientists have expressed concern that the concept-oriented calculus (so-called "reform" calculus) leaves students ill prepared for the rigors of real science where differential equations need to be solved, not just understood. This has led to some divergence of advice and enrollments in which mainstream ("real") calculus remains focused on complex techniques while reform calculus (and its reformed high school prerequisites) is recommended for "weaker" students who are not likely to ever make use of calculus. A similar caste system may emerge as QL-like conceptual statistics courses take root.

If this worry proves accurate, it may be difficult to prevent QL from becoming known as a lower track. Since neither traditional nor reformed calculus does very much to enhance QL, it is likely that statistics may have a greater impact than calculus on the social status of QL on campuses. But if even the traditional techniques-oriented Statistics 101 has lower status than Calculus 101, what will become of QL if it is identified with a conceptual Statistics 101 that makes no claim to provide research techniques?

Insider: The question is (or should be) "What do we want a 'QL certified' student to know and be able to do?" If the answer is "real research" then Calculus 101 won't meet the standard, nor will much of anything else at the undergraduate level. If we want the student to be able to understand and think critically about the physical, social, and political world around us, then Statistics 101 might be seen as inferior to a concepts oriented statistics course. Certainly that is true if by Statistics 101 we mean the traditional course that is offered on many campuses. However, a good Statistics 101 course can, like a concepts oriented "stat and QL" course, go a long way toward helping the student develop critical thinking skills.

Outsider: Many people outside the mathematics and statistics community argue that for QL to succeed, it must be embedded in courses across the curriculum (to reinforce foundation courses taught by mathematicians or statisticians). Many others worry that faculty in other fields suffer from their own versions of "math anxiety," so they are unlikely to be effective allies in spreading QL across the curriculum. Arguably, even more faculty may suffer from "stats anxiety," or even more likely, from statistical ignorance. After all, they too suffered through the infamous Statistics 101 as college students.

On the other hand, unlike mathematics, statistics is taught in departments across the campus, and many departments have faculty who identify themselves professionally as statisticians. So in theory, statistics could serve as a vehicle for QL across the curriculum. But in practice, statistics in these outlying departments is often marginalized to the courses taught by the statistics specialist. This, of course, would defeat the goal of "QL across the curriculum." What is the likelihood that the statistical diaspora could influence their host departments sufficiently to make "across the curriculum" a natural happening?

Insider: Unfortunately, this is not very likely. As an ideal, it sounds good. In practice, the reality is as you say: statistics is often marginalized in outlying departments. One problem is that outlying departments tend to see "statistics" as a collection of specific numerical techniques used by researchers in their fields, rather than as the broad and deep discipline that statistics truly is. Few faculty scattered across campus would identify mathematics as being just about Abelian groups, for example, but they tend to think of statistics as being just that slice of it that they regularly see. With its emphasis on inference techniques, the traditional Statistics 101 course has reinforced this narrow view. "Reform statistics," with its emphasis on data collection, graphics, and the like, pushes in the opposite direction. But there is a tremendous amount of inertia to be overcome. If a change is to be seen in how outlying departments use statistics, and enrich the QL component of their work, it most likely will need to depend on the broadening of views of statistics held by their faculty.

Examples of Quantitative Literacy

Whenever educators develop courses and tests in well-established subjects such as statistics or calculus, the central part of the subject is generally prescribed by tradition and common agreement: elementary statistics includes probability, correlation, and tests of hypotheses; calculus includes derivatives, integrals, their relationship and applications. Quantitative literacy has no such canon. Instructors apply their energy and imagination as they see fit to introduce students to examples of quantitative thinking that they find to be important and valuable. The result is a vast array of examples that represent both the breadth of quantitative needs of contemporary life as well as the variety of educational expectations for QL.

This section offers a few examples of this variety, adapted from different institutions, courses, and assessments. It is by no means a representative sample, merely one that illustrates some different strands of thinking by those who are beginning to focus seriously on QL expectations for students. (Specific sources are not cited both because some of these examples are currently in use and because they have been freely adapted to suit their use in this section.)

Numerical Common Sense

1. What is the population of the world?
   a) 40,000,000 b) 300,000,000 c) 6,000,000,000 d) 12,000,000,000
2. How many people live in the US?
   a) 40,000,000 b) 300,000,000 c) 6,000,000,000 d) 12,000,000,000
3. The average house price in the United States is closest to which number?
   a) $75,000 b) $120,000 c) $185,000 d) $420,000

Reading Graphs

1. The graph below shows data on the number of cells in a culture measured at six different times. Approximately how many cells are present after 6 hours?

a) 10,000
b) 6,000
c) 1,000
d) 5,000
e) 8,000

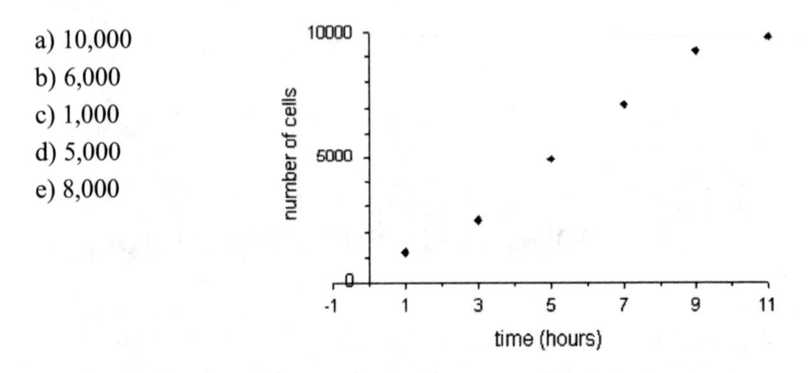

2. The graph below shows the percent change in the value of a company's stock:

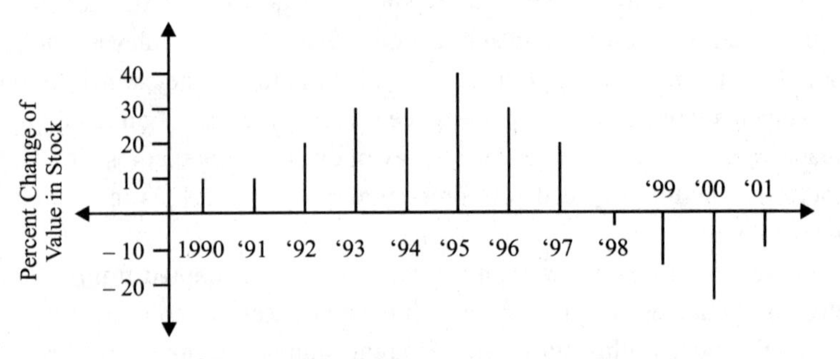

From the information in this graph, what appears to have occurred?
a) The value of the stock reached its highest value in 1995.
b) The value of the stock was still increasing in 1995, but began to decline the next year.
c) The value of the stock increased in 1995 and the value also increased but at a slower rate in the next year.
d) There is not enough information provided to answer the question.

3. The graph on the right shows enzyme activity as a function of temperature for a typical enzyme:

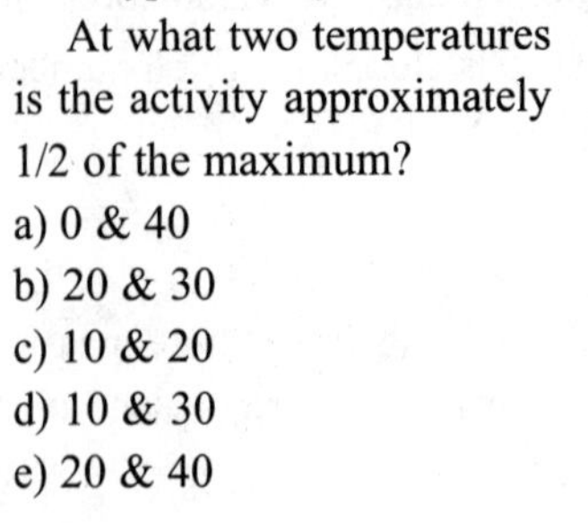

At what two temperatures is the activity approximately 1/2 of the maximum?
a) 0 & 40
b) 20 & 30
c) 10 & 20
d) 10 & 30
e) 20 & 40

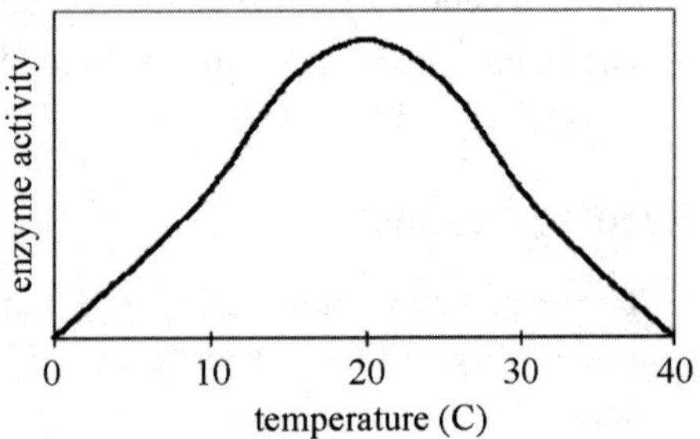

4. Inequality in distribution of income is usually illustrated by extremes: 45% of the income in the U.S. goes to the top 20% of households, whereas only 5% goes to the lowest 20%. A more thorough way to portray the same information is through a Lorenz curve that plots the percent of income in relation to the percent of households. The area between the Lorenz curve and the 45-degree line of perfect equality is called the Gini coefficient, which serves as a measure of the inequality of income in a society.

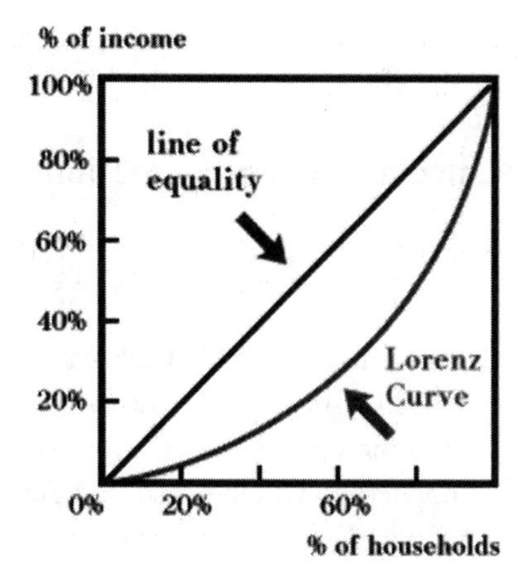

A graph that displays the Gini coefficient for different nations in relation to their per capita gross domestic products shows the unique position of the United States:

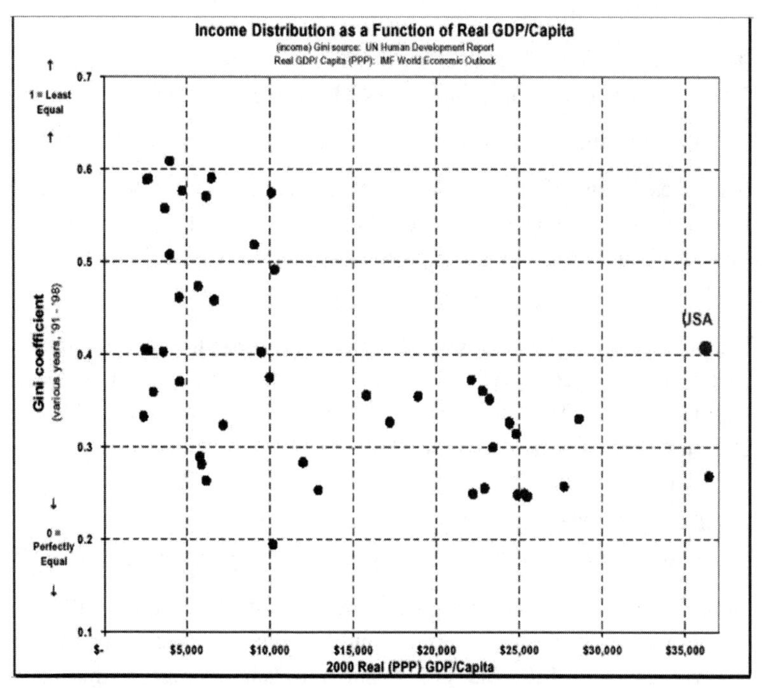

Describe in several sentences the characteristics of the countries in the three clusters evident in this graph—those in the lower left corner, those above them in the upper left quadrant, and those in the lower right part of the graph.

Calculation and Estimation

1. What is an approximate 15% tip on a restaurant bill of $62.84?

   a) $6.30 b) $9.50 c) $3.15 d) $12.00

2. According to the 2000 census, the population of Lincoln county is approximately 100,000. A power company predicts that the county's population will increase by 7% per year while the county supervisors predict that the population will increase by 7,500 each year. Which group predicts the larger population in 10 years?

   a) Power company

   b) Supervisors

   c) Both predictions are the same after 10 years

   d) There is not enough information provided to answer the question.

3. Comparing 8 with 2, is 8

   (a) four times as much as 2; (b) 300% more than 2;

   (c) four times more than 2; (d) 400% more than 2?

   Is 2 (a) four times less than 8; (b) 75% less than 8; or (c) 25% of 8?

4. If it takes 5 cans of spray paint to cover a garden sculpture that stands 3 feet tall, how many cans would it take to cover an enlarged version of the same sculpture (with the same proportions) that stands 6 feet tall?

   a) 10 b) 15 c) 20 d) 25

5. The percentage of students in a school who passed an important standardized test decreased by 15% from 2000 to 2001. After worried teachers redoubled their efforts, the percentage who passed increased by 15% from 2001 to 2002. In which of these years was the percentage of students who passed the highest?

   a) 2000 b) 2001 c) 2002

   d) There is not enough information provided to answer the question.

6. Sarah is an avid bowler with a game average of 124. She typically bowls five games on league night at the bowling alley. One week her first three games were 117, 130, and 113. What must she average in the last two games to achieve her overall average of 124?

   a) 130 b) 124 c) 120 d) 134 e) 129

7. An "A" tent (that is open in the front and back and has no floor) has an opening that is 12 ft wide. The tent is 10 ft long and 8 ft high. What are the dimensions of the tarp used for the covering of the tent?

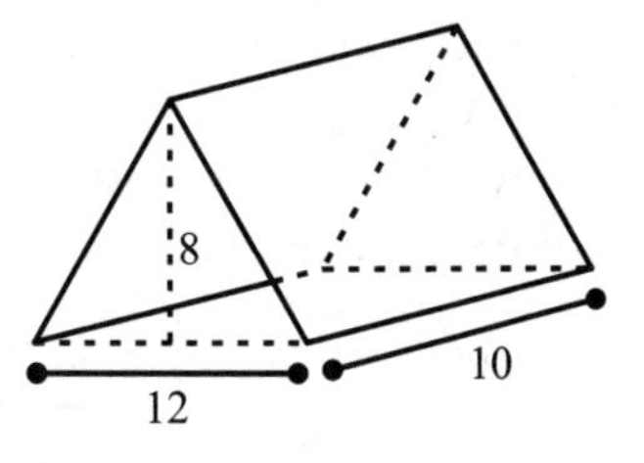

   a) 10 ft × 20 ft b) 6 ft × 8 ft
   c) 10 ft × 10 ft d) 12 ft × 12 ft

Understanding Percentages

1. When a company changed the way it handled its employee retirement plan from a positive to a negative check off, the percentage of employees who joined the plan jumped from 30 to 90 percent. (In a positive check-off system, employees must apply to participate; in a negative checkoff system, employees are automatically enrolled unless they actively decline to be part of the program.) In reporting on this change, proud company officials described it as resulting in a 200% increase in participation, but journalists said that it was only a 60% increase. Which is correct?

2. The incidence of breast cancer in women over 50 is approximately 380 per 100,000. If cancer is present, a mammogram will detect it approximately 85% of the time and miss it approximately 15% of the time; however, if cancer is not present, the mammogram will indicate cancer approximately 7% of the time. Statisticians call 15% the "false negative" rate (falsely certifying the absence of cancer) and the 7% the "false positive" rate (falsely indicating the presence of cancer). From these percentages it appears that the false positive rate for mammograms (for women over 50) is about half that of the false negative rate.

   However, if these percentages are translated into actual numbers of women, the picture looks quite different:

	Cancer Present	Cancer Absent	Totals
Mammogram:			
Signals Cancer	323 (85%)	6,677 (7%)	7,000
Signals All-Clear	57 (15%)	92,943 (93%)	93,000
Totals	380	99,620	100,000

   What do these data tell us? Is the risk of a false positive higher or lower than the risk of a false negative? Which mistake (false result) is more serious? What do these data imply for recommendations and public policy concerning mammogram screening?

3. The following table shows 1997 unemployment rates by race and educational background. Describe in words the meaning of the number 6.1 shown in the next to last column of the Black row. What is the significance of this number in the context of other entries in this table?

1997 US Unemployment Rates by Educational Attainment & Race					
(Percent unemployed of the civilian labor force)					
		—Highest Grade Achieved—			
RACE	**Total**	**<12**	**12**	**13–15**	**>15**
Total:	4.4	10.4	5.1	3.8	2.0
White:	3.9	9.4	4.6	3.4	1.8
Black:	8.1	16.6	8.2	6.1	4.4
Other:	7.3	9.6	7.5	5.5	3.0

4. One of the major health stories of 2002 was the decision by NIH to end early a major clinical trial of hormone replacement therapy (HRT) due to increased risks of stroke and breast cancer. The day the HRT decision was announced, headlines, news articles, and the NIH press briefing itself all emphasized a 26% higher risk of breast cancer, a 29% increase in heart attacks, and a 41% higher risk of strokes among women taking the hormone therapy.

   Virtually none of the early news accounts said what the actual risks were. What they did say, for example, is that of 10,000 women taking the combined therapy as compared with the same number taking a placebo, 8

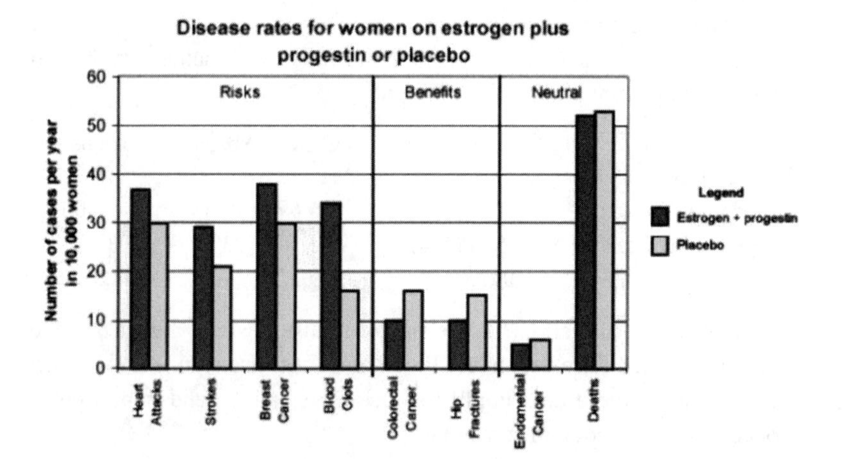

more will have strokes each year. So what does the 41% increase in risk of stroke for women taking HRT actually mean? What is the risk for women not on HRT as compared with those who are? Is the increase as great as 41% may suggest? (Data on this and other risks can be found on the accompanying graph.)

Reading a Newspaper

For each of the following excerpts, please explain briefly whether you believe the data and the conclusions the author has reached, your reasons for believing or not, and identify what other information you would want to have in order to better assess the analysis contained in the excerpt.

1. **Only 7% Seek to Transfer to a Better School**
Despite U.S. Plan, Most City Pupils Not About to Move

Stephanie Banchero and Ana Beatriz Cholo, *Chicago Tribune*, August 20, 2002

Whether it's because parents are satisfied with current conditions or were provided few good alternatives, only 1,900 Chicago Public Schools pupils—about 7 percent of those eligible—have requested transfers out of failing schools, district officials said Monday as they tallied the last applications. School officials will spend the next few days trying to match parents' wishes with the estimated 2,900 available classroom slots, giving priority to the lowest achieving low-income pupils.

Though it was unknown Monday how many of the 1,900 youngsters would ultimately be moved, officials said they hoped to let parents know by Friday whether there is space for their children in a nearby, better performing school. Under a new federal law, about 29,000 Chicago students were offered the opportunity to flee their under-performing school and be bused to a better one. By Monday it was clear that the vast majority of parents prefer to stay in neighborhood schools, whether or not they are failing.

2. **New Intel Chip Promises to Do More With Less**

By Bloomberg News, *New York Times,* August 14, 2002

The Intel Corporation said today that its next personal computer chip would pack in twice as many transistors as the company's best-selling Pentium 4, use less power and have some of the tiniest parts ever made in high-volume chip manufacturing.

Intel, the world's biggest semiconductor maker, revealed details of four manufacturing changes being used to produce the chip, which is code-named Prescott and expected in the second half of 2003. The design will eventually appear in everything from powerful servers that run corporate networks to small hand-held devices.

As sales fall, Intel and its rivals are building faster chips to spur demand. Intel made memory chips in March with the new methods, which are being perfected in an Oregon plant. The circuit size was cut to 90 nanometers; 50-nanometer gates to conduct electricity are the thinnest ever built, Intel said. A human hair is about 2,000 times as wide.

3. **Don't Bet on Bipartisan Niceties**

David S. Broder, *Washington Post*, January 1, 2003

Each year Congressional Quarterly counts the number of votes on which a majority of Republicans oppose the stand taken by a majority of Democrats. Then it calculates the percentage of times on which each member has voted with the party majority on those roll calls.

When I averaged the year-by-year results for both chambers, I found the percentage of partisan-divide roll calls has gone from 39 percent in the 1970s to 47 percent in the 1980s to 58 percent in the 1990s. Even more striking is the growth in cohesion—call it discipline or philosophical agreement—within both party caucuses. In the 1970s, on the partisan roll calls, the average member of Congress backed the party position 65 percent of the time. In the 1980s, the average degree of partisan loyalty rose to 73 percent; and in the 1990s, to 81 percent. In these past two years, it has been 87 percent.

Another way of expressing that trend is this: In the 1970s, there were no years in which the average Republican or Democrat voted "right" from a partisan view as much as 75 percent of the time. In the 1990s, by contrast, there was only one year in which the Republican average fell as low as 74 percent—and none in which the Democratic average was below 79 percent.

4. **ACT test scores dip slightly**
Drop due to younger students taking exams

CNN, August 21, 2002

Scores dipped for the high school class of 2002 on the ACT college entrance exam, breaking a five-year streak during which results remained unchanged, the test maker said Wednesday.

ACT Inc., a non-profit based in Iowa City, Iowa, said the drop was to be expected because Illinois and Colorado began requiring that all high school juniors, starting with the class of '02, take the test whether or not they are enrolled in college-prep courses. Using scores students received on the latest ACT they took, 2002 graduates averaged a composite of 20.8, down from the 21 average maintained from 1997 to 2001.

More students than ever took the ACT, 1.12 million of this year's graduates, or about 46,000 more than last year. This year's results are the first since Illinois and Colorado began assessing their public schools by having all 11th-graders take the ACT at state expense, even if they don't plan on attending college. Richard Ferguson, chief executive of ACT, said in a statement that the requirement had a positive result: "Thousands of students in Illinois and Colorado who had not indicated an interest in attending college were identified as ready for college coursework."

5. **The Calculator. How Kenneth Feinberg Determines the Values of Three Thousand Lives**

Elizabeth Kolbert. *The New Yorker*, Nov. 25, 2002.

At first glance, the tables [for the Victims' Compensation Fund] defy most notions of equity; the more needs a family is likely to have, the less well it fares. For example, the tables show that the widow of a twenty-five-year-old who had no children and was earning a hundred and twenty-five thousand dollars a year can anticipate a payment, before any offsets, of nearly four and a half million dollars. The widow of a man who was earning the same salary and was similarly childless but was forty can expect half that amount, while the widow of a forty-year-old who was making fifty thousand dollars and had one child can expect a quarter of it. Finally, the widow of a forty-year-old with two dependent children who was making twenty thousand dollars does worst of all. She can expect about a fifth, or slightly more than nine hundred thousand dollars.

Common Conundrums

1. The city of Nicelife, like many towns across America, has a growing population of poor families who are unable to fully provide for their own needs. Children from these families participate in the federal school lunch program, which provides educational leaders an unobtrusive way to identify children who are at risk from the educational deficits known to accompany poverty. As required by state law, and because they are committed to ensure good education for all students, the Nicelife school district disaggregates scores on state tests into two groups that are reported separately: school lunch students, and non-school lunch students. Scores on state tests are widely publicized for each school, and for each group of students within each school.

Last year, parents and teachers in Nicelife was very pleased because its average scores on the required 8th grade mathematics test exceeded the state average for both groups of students:

	Average 8th Math Score	
	Nicelife	State
Non-School Lunch Students	85	80
School Lunch Students	60	50

Imagine the dismay when they found Nicelife on a list published in the state's leading newspaper of school districts whose mathematics scores were below the state average:

	Average 8th Math Score	
	Nicelife	State
Niceville	72.5	77

Many citizens were outraged, but then they realized that it is just the way averages work: half of Niceville's students are in the school lunch program, whereas only 10% of the state's students are.

(On January 8, 2002, President Bush signed into law a reform of the Elementary and Secondary Education Act (ESEA) that requires annual testing for all students in grades 3–8 with results broken out by poverty, race, ethnicity, disability, and limited English proficiency. Very soon the public will become aware of many instances of the apparent paradox that Nicelife discovered last year.)

2. In January 2001, well before Al Gore decided not to be a candidate for president in 2004, New York Times columnist William Safire reported the following odds for potential Democratic candidates for the presidential nominations of 2004:

Tom Daschle	4-1	Pat Leahy	6-1
Joe Biden	5-1	Joe Lieberman	5-1
Richard Gephardt	15-1	Chris Dodd	4-1
John Edwards	9-1	Russell Feingold	8-1
John Kerry	4-1	Al Gore	2-1

What are we to make of this? If five candidates all have equal chances of winning an election, then the odds against any one of them will be 4-1. This seems to be the status of Tom Daschle and John Kerry in Safire's report. But then Al Gore, at 2-1 odds, seems distinctly better off since his chances in this field of ten are the same as if he had only two equally strong opponents. By the time you add in Joe Biden and Joe Lieberman—both nearly as strong as Daschle and Dodd, the probabilities of victory quickly add up to more than one. Assuming that at the time of his writing Safire was correct about each candidate's relative strengths, what were their real odds?

National Numeracy Network

Among the dozens of colleges and universities who are exploring ways to enhance students' quantitative literacy, those listed below have served as the core of what came to be called the *National Numeracy Network*. Each focuses on a different aspect of quantitative literacy; collectively they serve as a resource for others.

Dartmouth College: Putting QL Material on the Web

The Center for Mathematics Education at Dartmouth is developing a web site that will contain an extensive collection of QL materials directly usable in the classroom. The Center is intended to serve as a repository of QL resources that benefit teachers across all levels and all disciplines. This effort grows out of an earlier NSF-supported project on Mathematics Across the Curriculum (MATC). While continuing to include interdisciplinary mathematics materials, the project is adding materials with a quantitative literacy orientation and making a special effort to include materials that address QL issues specifically for K–12 teachers and first- and second-year college students. Wherever possible, electronic materials will be made available directly from the site, but where this is not possible the site will provide information on materials and how to obtain them.

The project is actively seeking QL materials from teachers at all levels that they have used successfully in their classes. These need not be new materials, nor of any particular length, but to satisfy some of the goals of quantitative literacy they must be more than merely mathematical in focus. The goal is to provide valuable resources to secondary and post-secondary educational institutions, departments, individual faculty members, students, as well as the general public.

To the extent that time and resources permit, the Dartmouth Center intends to support QL workshops at other locations with materials, connections with presenters, and workshop design. Through such activities it hopes to help new projects get started and to crystallize public understanding of QL through concrete usable materials. *Contacts*: Kim Rheinlander, Dorothy Wallace; URL: http://www.math.dartmouth.edu/~matc/.

Hollins University: Quantitative Reasoning Across the Curriculum

In 1998 Hollins University, a small liberal arts university for women in Roanoke, VA, implemented a basic skills quantitative reasoning requirement to ensure that all graduates have an understanding of quantitative methods. Three years later the program was expanded to include both a basic skills and an applied skills requirement. The goal of the applied skills requirement is for all Hollins graduates to appreciate how quantitative skills apply to their chosen field of study in the liberal arts curriculum.

To help implement this requirement, Hollins sought and received support of the National Science Foundation to adapt ideas and materials from Mathematics across the Curriculum (MATC) programs at other institutions. This Faculty Development Program for Quantitative Reasoning brings scholars to campus to lead interactive workshops for faculty across the campus who are creating QL courses by adapting and modifying the ideas of these scholars. The goal of this project, which was inspired by the 1996 report entitled "Quantitative Reasoning for College Graduates" of the MAA Committee on the Undergraduate Program in Mathematics (CUPM), is to develop at Hollins a commitment to QL across the curriculum that is securely rooted in faculty professional development. *Contact:* Caren Diefenderfer.

Macalester College: Quantitative Methods for Public Policy

Macalester College in St. Paul, MN, is a liberal arts college of 1800 students with a long-standing emphasis on public policy. Yet despite the fact that half of all recent Macalester graduates have studied statistics and over two-thirds have taken some mathematics course, a recent survey revealed that large segment of Macalester students believe that "quantitative methods are irrelevant to their lives and their interests." Faced with this QL challenge, Macalester created a new cross-disciplinary program predicated on the premise that all Macalester students should to be able to read critically any article with quantitative content in the *New York Times*.

With this goal in mind, a faculty design team concluded that to be successful, a QL program must not only be housed within the client disciplines but must be part of the mainstream courses in these disciplines. Since literacy and discourse are tightly interwoven, the design team argued that QL would thrive only in an environment where students routinely exercise skills of communication and discourse. For this to happen, the QL experience should be shared simultaneously by a large fraction of the student body and not isolated in individual, independent courses.

To achieve these objectives, Macalester enlisted faculty from several disciplines who teach courses that bear in some way on public policy to agree on a common quantitatively rich focus that can be used as a source of motivation and examples. For the first year (2002–03), conceived as a pilot study, they chose as their common focus the topic of school vouchers; for the second year (2003–04) the focus is immigration policy. Following a year of planning, and several months before the first classes would be held, participating faculty held a workshop in which they presented to each other the different perspectives on QL from the viewpoint of their specific disciplines. Based on the results of this pilot year, Macalester received a three-year FIPSE grant to continue the project on their own campus and to run workshops for faculty at other colleges who may be thinking along similar lines. *Contact:* David Bressoud.

Trinity College: Expanding the Reach of QL Materials

The Mathematics Center at Trinity College in Hartford, CT, is engaged in several activities to enlarge an established tradition of their own work in quantitative literacy both on and beyond the Trinity campus. These include:

- *Outreach to the greater Hartford Schools.* Trinity is coordinating workshops at the Greater Hartford Academy of Mathematics and Science (adjacent to the Trinity campus) as part of their day-long programs for teachers in the greater Hartford area. The QL workshops focus on incorporating real world data into high-school courses.
- *Producing materials for the Dartmouth QL website.* The Mathematics Center assists Trinity faculty who teach existing QL courses and labs to make their materials coherent, attractive, and transferable. Initial courses involved in this dissemination are *Contemporary Applications: Mathematics for the 21st Century, Logic in the Media,* and *Skepticism and Belief.*
- *Course Development.* The Mathematics Center recruits and supports mathematics consultants to assist faculty members in disciplines other than mathematics who are developing new courses that incorporate quantitative reasoning. It also supports internships for "Math Associates," TAs chosen from among the Center's Tutors to work with faculty during the summer and help with these new QL courses.

- *Cooperation with the Washington Center.* Trinity Center personnel are preparing workshops in conjunction with the Washington Center that focus on establishing QL programs at institutions across the country.

These activities and materials are disseminated also through the Northeast Consortium for Quantitative Literacy (NECQL). *Contact*: Judith Moran.

University of Nevada, Reno: Bringing QL from Media into the Classroom

The Numeracy Center at the University of Nevada, Reno is producing a series of QL modules for use in grades in 7–14. Each module contains one or two examples from the popular media, an analysis of its relevance to appropriate grade levels, and suggestions for how to incorporate it into a class. The focus is on classes or contexts such as science, social science, or history. The intention is for the modules also to be used in teacher workshops.

Modules are prepared by a team of 4–5 faculty from the Department of Mathematics in consultation with a dozen or so faculty from other departments who will offer suggestions for topics and critique drafts. There is a cadre of faculty and a culture from UNR's former Math Across the Curriculum project (funded by NSF) that supports this QL work. Within a year or two they expect to have ten modules at various grade levels 7–12, plus five for undergraduate grades 13-14. Modules currently being developed include:

- *Magnitudes: How Big is A Billion?* We hear about billions of dollars being spent by governments. How big is a billion? How much do a billion dollar bills weigh? How high would they be if stacked on one another? Similar entertaining and instructive ideas can be exploited to convey some feeling for large numbers.
- *Consumers Need Numeracy.* Examples: Which is the better buy, a large or small pizza? A two-liter bottle of soft drink or a six pack? Buying carpet by the square yard or square foot. It happens with some frequency that a larger size is a worse buy than a smaller size.
- *Better Nutrition through Numeracy*. Percents and graphs in developing a better understanding of proper nutrition; e.g., reading food labels on packaged foods and nutrition charts available at many fast food restaurants.
- *Numeracy and Driving*. The Nevada Driver's Manual has several references to stopping times, stopping distances, and reaction times. Included are bar graphs and figures about time saved by speeding.

- *Weighted and Aggregated Averages.* A central example is apparent discrimination in hiring due to use of aggregate averages. Closer examination reveals that it may not be so.

Other tentative subjects include genetics and disease, DNA and forensics, and global warming. *Contact*: Jerry Johnson.

Washington Center: Creating QL Professional Development Opportunities

Washington Center for Improving the Quality of Undergraduate Education focuses on developing professional development opportunities in QL. Located at Evergreen State College in Olympia, Washington, the Center seeks to create professional development experiences and opportunities for faculty from two- and four-year colleges to learn about QL and incorporate it into the curriculum of their courses. It convenes meetings that involve cross-sector groups in discussions about expectations for students' learning in mathematics and QL in the Washington state K–16 system; prepares articles for web and print distribution about the practice of QL across the curriculum as well as content analyses of expectations for students' learning of mathematics and QL throughout the K–16 system; and it makes explicit connections between QL and Washington Center's initiatives focused on increasing the academic success of under-represented students, including students of color, in both two-and four-year colleges.

Materials generated through these workshops are sent to Dartmouth for posting on their web site, as are assessments of students' learning. These also include reflective papers from a symposium for mathematics educators, mathematicians, multicultural directors, and faculty involved in both mathematics across the curriculum and quantitative literacy to discuss the connections and disconnections between reforming the mathematics curriculum and developing curriculum to increase students' quantitative literacy.

Participants in Washington Center projects are asked to design courses around an *organizing question* rather than around a set of disciplinary principles or a smorgasbord of QL techniques. This approach inverts traditional course design in that the organizing question is the focus of curricular design, not the listing of topics that need to be included. *Contacts:* Rob Cole, Emily Decker, Gillies Malnarich

Acknowledgements, Authors, Advisors

(Affiliations as of the original meetings)

Mathematics and Democracy Design Team

GAIL BURRILL, Director, Mathematical Sciences Education Board

SUSAN L. GANTER, Department of Mathematical Sciences, Clemson University

DANIEL L. GOROFF, Associate Director, Derek Bok Center, Harvard University

FREDERICK P. GREENLEAF, Professor of Mathematics, Courant Institute, New York University

W. NORTON GRUBB, Graduate School of Education, University of California-Berkeley

JERRY JOHNSON, Professor of Mathematics, University of Nevada, Reno

SHIRLEY M. MALCOM, Head, Directorate for Education and Human Resources Programs, American Association for the Advancement of Science

VERONICA MEEKS, Mathematics Teacher, Western Hills High School, Fort Worth, TX

JUDITH MORAN, Director, Mathematics Center, Trinity College, Hartford, CT

ARNOLD PACKER, Chair, SCANS 2000 Center, Johns Hopkins University

JANET P. RAY, Professor of Mathematics, Seattle Central Community College, Seattle, WA

C. J. SHROLL, Executive Director, Workforce Development Initiative, Michigan Community College Association

EDWARD A. SILVER, Department of Mathematics, School of Education, University of Michigan

LYNN A. STEEN, Professor of Mathematics, St. Olaf College, Northfield, MN

JESSICA UTTS, Professor of Statistics, University of California-Davis

DOROTHY WALLACE, Professor of Mathematics, Dartmouth College, Hanover, NH

National Numeracy Steering Committee

ROBERT COLE, Professor of Mathematics, Evergreen State University, Olympia, WA

EMILY DECKER, Director, Washington Center for Improving the Quality of Undergraduate Education

JERRY JOHNSON, Professor of Mathematics, University of Nevada, Reno

RAYMOND JOHNSON, Professor of Mathematics, University of Maryland

LINDA KIME, Mathematics Department, University of Massachusetts, Boston

RICHARD MILLER, Office of Education and Quality Initiatives Assoc. of American Colleges and Univ.

JUDITH MORAN, Director, Mathematics Center, Trinity College, Hartford, CT

JANET P. RAY, Professor of Mathematics, Seattle Central Community College, Seattle, WA

LINDA ROSEN, Senior Vice President for Education, National Alliance of Business

HENRY (LEN) VACHER, Professor of Geology, University of South Florida

DOROTHY WALLACE, Professor of Mathematics, Dartmouth College, Hanover, NH

Endnotes

1. Lynn Arthur Steen (editor). *Mathematics and Democracy*. Princeton, NJ: National Council on Education and the Disciplines, 2001.
2. Bernard L. Madison and Lynn Arthur Steen (editors). *Quantitative Literacy: Why Numeracy Matters for Schools and Colleges.* Princeton, NJ: National Council on Education and the Disciplines, 2003.
3. Alan Schoenfeld. "A Convergence of Needs." In *Why Numbers Count: Quantitative Literacy for Tomorrow's America.* Lynn A. Steen, editor. New York, NY: The College Board, 1997.
4. National Center for Educational Statistics. *Digest of Education Statistics 2001.* Washington, DC: U. S. Department of Education, 2001.
5. Thomas D. Snyder, Editor. *120 Years of American Education: A Statistical Portrait.* Washington, DC: U. S. Department of Education, Office of Educational Research and Improvement, January 1993.
6. Snyder, *ibid.*
7. Diane Ravitch and Joseph Viteritti. *Making Good Citizens: Education and Civil Society.* New Haven, Yale University Press, 2001.
8. One example is the International Statistical Literacy Project (formerly called the World Numeracy Project). URL: course1.winona.edu/cblumberg/islphome.htm.
9. Judith Ramaley, *et al. Greater Expectations.* Washington, DC: Association of American Colleges and Universities, 2002.
10. "Programs that Enhance Learning," found in *America's Best Colleges, US News and World Report.*
11. George Kuh, private correspondence.
12. John Allen Paulos. *Innumeracy: Mathematical Illiteracy and its Consequences.* New York, NY: Hill and Wang, 1988.
13. Frank Newman, *et al. The Futures Project.* URL: www.futuresproject.org
14. Clifford Adelman. "The Empirical Core." *University Business,* 2 (April 1999) 46–47.
15. Stephen P. White, et al. *The New Liberal Arts.* New York, NY: Alfred P. Sloan Foundation, 1981.

16. PISA: Program for International Student Assessment. URL: www.pisa.oecd.org/math/def.htm
17. Mathematics Learning Study Committee. *Adding It Up*. Washington, DC: National Academy Press, 2001, p. 5.
18. Theodore M. Porter. *Trust in Numbers*. Princeton University Press, Princeton, NJ, 1996.
19. Peter Bernstein. *Against the Gods: The Remarkable Story of Risk*. New York, NY: John Wiley & Sons, 1996.
20. Michael Meyerson. *Political Numeracy: Mathematical Perspectives on Our Chaotic Constitution*. New York: W. W. Norton & Co., 2002.
21. Steve Kennedy and Deanna Haunsperger, "Math Makes the Movies." *Math Horizons*, November 2001, p. 5–10.
22. Board on Life Sciences. *BIO 2010: Transforming Undergraduate Education for Future Research Biologists*. Washington, DC: National Academy Press, 2002.
23. Patricia Cline Cohen. *A Calculating People: The Spread of Numeracy in Early America*. Chicago, Ill, University of Chicago Press, 1982; New York: Routledge, 1999.
24. Richard W. Clark. *Dual Credit: A Report of Programs and Policies that Offer High School Students College Credit*. Pew Charitable Trusts, 2001.
25. Andrew Sum, Irwin Kirsch and Robert Taggart. *The Twin Challenges of Mediocrity and Inequality: Literacy in the U. S. from an International Perspective*. Princeton, NJ: Educational Testing Service, 2002, p. 3.
26. Milton Goldberg and Susan L. Traiman. "Why Business Backs Education Standards." In Diane Ravitch, ed., *Brookings Papers on Education Policy*. Washington, DC: Brookings Institution, 2001.
27. Sum *et al. op. cit.*, pp. 9, 15.
28. Sum *et al. op. cit.*, pp. 24–25.
29. Office of Economic Cooperation and Development (OECD). *Literacy in the Information Age: Final Report of the International Adult Literacy Survey*. Quebec: Statistics Canada, 2000.
30. Richard Rothstein. "SAT Scores Aren't Up. Not Bad, Not Bad at All." *New York Times*, August 29, 2001.
31. Richard Pérez-Peña. "Court Reverses Finance Ruling on City Schools." *New York Times,* June 26, 2002.
32. Andrew Sum. *Literacy in the Labor Force: Results from the National Adult Literacy Survey*. Washington, DC, National Center for Educational Statistics, U. S. Department of Education, 1999.
33. Richard Judy and Carol D'Amico. *Workforce 2020: Work and Workers in the 21st Century*. Indianapolis, IN: Hudson Institute, 1997.

34. Sum et *al. op. cit.*, p. 29.
35. David J. Lutzer, James W. Maxwell, and Stephan B. Rodi. *Statistical Abstract of Undergraduate Programs in the Mathemat-ical Sciences in the United States: Fall 2000 CBMS Survey*. Providence, RI: American Mathematical Society, 2002. URL: www.ams.org/cbms/
36. Laurie Buxton. *Math Panic*. Portsmouth, NH: Heinemann Educational Books, 1991.
37. Linda Sons, *et al. Quantitative Reasoning for College Graduates: Report of a CUPM Subcommittee on Quantitative Literacy Requirements*. Washington, DC: Mathematical Association of America, 1996. URL: www. valpo.edu/home/faculty/rgillman/ql/
38. Alan Tucker, *et al. The Mathematical Education of Teachers*. Conference Board of Mathematical Sciences, Issues in Mathematics Education, Vol. 11. Providence, RI: American Mathematical Society, 2001. URL: www. cbmsweb.org/MET_Document/index.htm
39. Reports from this "Curriculum Foundations" project are available online at www.maa.org/features/currfound.html.
40. John D. Bransford, Ann L. Brown, and Rodney R. Cocking, (Editors). *How People Learn: Brain, Mind, Experience, and School.* Washington, DC: National Academy Press, 1999.
41. Nancy Baxter Hastings (editor). *Rethinking the Courses Below Calculus.* Washington, DC: Mathematical Association of America, (forthcoming). See also Arnold Packer, "College Algebra." URL: www.maa.org/t_and_l/ college_algebra.html
42. A similar effort at the high school level was developed in the 1990s by the College Board under the name "Pacesetter."
43. Committee on the Undergraduate Program in Mathematics. *Universal Mathematics, Parts I and II.* Washington, DC: Mathematical Association of America, 1954, 1958.
44. John G. Kemeny, J. Laurie Snell, and Gerald Luther Thompson. *Introduction to Finite Mathematics*. Englewood Cliffs, NJ: Prentice-Hall, 1957.
45. Consortium on Mathematics and its Applications (COMAP). *For All Practical Purposes: Introduction to Contemporary Mathematics*. New York : W.H. Freeman, 1988.
46. Paul E. Barton, *Meeting the Need.* Princeton, NJ: Educational Testing Service, 2002.
47. John Sinclair. *The Statistical Account of Scotland.* Edinburgh, 1798, vol. 20, xiii.

48. See "statistick" in John Walker, *A Critical Pronouncing Dictionary and Expositor of the English Language*. Philadelphia, 1803; and "statistics" in Noah Webster, *A Compendious Dictionary of the English Language*. New Haven, 1806.
49. This movement was led by the American Statistical Association (ASA) and the National Council of Teachers of Mathematics (NCTM) with funding from the National Science Foundation and several private sources.
50. National Council of Teachers of Mathematics. *Principles and Standards for School Mathematics.* Reston, VA: NCTM, 2000.
51. Richard Feynman. *Surely You're Joking, Mr. Feynman!* New York, NY: W.W. Norton & Co., 1985.
52. Howard Gardner. *The Unschooled Mind: How Children Think and How Schools Should Teach.* New York, NY: Basic Books. 1991, p. 165.
53. Richard W. Clark, *op. cit.*
54. The Committee on Articulation and Placement (CAP), created in 2001, is a standing committee of the Mathematical Association of America that has the presidents of AMATYC and NCTM as ex-officio members.
55. David Lutzer, *et al. op. cit.*
56. Richard W. Clark, *op. cit.*
57. *Access to Excellence*, A Report of the Commission on the Future of the Advanced Placement Program, College Board, 2001. New York, NY: The College Board, 2001.
58. Committee on Programs for Advanced Study of Mathematics and Science in American High Schools, National Research Council. *Learning and Understanding: Improving Advanced Study of Mathematics and Science in U. S. High Schools.* Washington, DC: National Academy Press, 2002.
59. Eugenio J Gonzalez, Kathleen M. O'Connor, and Julie A. Miles. *How Well Do Advanced Placement Students Perform on the TIMSS Advanced Mathematics and Physic Tests?,* International Study Center, Boston College, 2001.
60. "Ticket to Nowhere," *Thinking K–16*, Vol. 3, Issue 2, Education Trust, 1999.
61. Bernard L. Madison. "An analysis of collegiate expectation statements about appropriate mathematics preparation for college mathematics." Preliminary report.
62. National Council of Teachers of Mathematics. *Standards Impact Research Group*. Abstracts available at: www.nctm.org/highered/researchbriefs.htm
63. Some sites are already available with preliminary or tangential material: Mathematics Across the Curriculum: hilbert.dartmouth.edu/~matc/index.html. Chance database: www.dartmouth.edu/~chance

64. Arthur Jaffe. "Ordering the Universe: The Role of Mathematics." In Edward E. David (editor). *Renewing U. S. Mathematics: Critical Resource for the Future.* Washington, DC: National Academy Press, 1984, p. 117–162.
65. William Thurston. "Mathematics Education." *Notices of the American Mathematical Society.* 37:7 (September, 1990) 844–850.
66. Bernard L. Madison and Therese A. Hart. *A Challenge of Numbers: People in the Mathematical Sciences*. Washington, DC: National Academy Press, 1990.
67. Alan Tucker, *et al. op. cit.*
68. James D. Watson, *The Double Helix*. New York, NY: Atheneum, 1968.
69. William Thurston, *op. cit.*